BARRON'

NEW JERSEY BIOLOGY COMPETENCY TEST

Cynthia Pfirrmann, M.S.
Scotch Plains-Fanwood High School
Scotch Plains, New Jersey

BARRON'S

© Copyright 2012 by Barron's Educational Series, Inc.

All inquires should be addressed to:
Barron's Educational Series, Inc.
250 Wireless Boulevard
Hauppauge, NY 11788
www.barronseduc.com

Library of Congress Catalog Card Number: 2012941191

ISBN: 978-0-7641-4378-6

Printed in the United States of America
9 8 7 6 5 4 3 2 1

10%
POST-CONSUMER
WASTE
Paper contains a minimum
of 10% post-consumer
waste (PCW). Paper used
in this book was derived
from certified, sustainable
forestlands.

Table of Contents

13 Human Body Systems / 203

14 Ecology / 223

15 Biomes / 241

16 Human Impacts / 255

Introduction

First year biology students in New Jersey public high schools are required to take the New Jersey Biology Competency Test (NJBCT). The test is given mid-May over a two day period throughout the state. The NJBCT's purpose is to assess whether you have mastered the biology content and skills that the New Jersey State Board of Education has identified as requirements for high school graduation. The general areas tested are based on the New Jersey Science Core Curriculum Content Standards (CCCS). Biology topics included in the current CCCS are heredity and reproduction, evolution and diversity, interdependence, organization and development, and matter and energy transformation.

The NJBCT is given over two consecutive days, with each session lasting one hour and forty-five minutes. If you are absent for one, or both, days of testing, you will be expected to complete the test during scheduled makeup days. Makeup testing is generally scheduled to take place over the two days immediately following the two main test days.

There are two types of questions presented on the NJBCT: multiple choice and performance assessment tasks (PATs). There are sixty multiple-choice questions and two PATs on the test. With two days to take the test, you can expect to be asked thirty multiple-choice questions and one PAT on each testing day. Multiple-choice questions will each have four answer choices (A, B, C, and D). You will be provided with an answer sheet to fill in the correct answer bubble for each question. Multiple-choice questions are each worth one point. PATS are based on real-life scenarios that present a biological problem or challenge. A PAT may include graphic material such as tables, maps, and graphs. In answering the PATs, you will be required to develop a solution and write a detailed explanation of how and why your chosen solution will work. The written answers to the PATs are read by at least two raters and graded against a scoring rubric. The raters' scores are averaged to give a final score between 3 and 12.

After all the tests are graded, you will receive a score report telling you your total score and your scores in specific content and skill areas. Total scores are rated as Advanced Proficient, Proficient, or Partially Proficient. In order to pass the NJBCT, you will need to achieve a total score of Advanced Proficient or Proficient. If you do not pass the test the first time, you will have other opportunities to demonstrate your mastery of the test content. Many schools offer extra help to students who score below the proficient level.

As a high school biology teacher, I'm frequently asked by students why they have to take this test and whether or not it really counts. This test has a historical basis in the federal No Child Left Behind laws. The purpose of these laws was to ensure that all public school students in the United Stated received at least a fundamental high school education by the time they were ready to graduate. Standardized tests have been developed in many states to gauge performance and proficiency in different subject areas. Different state strategies and new laws regarding educational policies may change some of the details over time, but standardization of content and testing seem to have become politically accepted mechanisms for evaluating student progress, teacher performance, school and district operations, and qualifications for state and federal funding. So, to answer the above posed questions, yes, public high school students have to take this test, because it is mandated by the state of New Jersey in order to evaluate and fund its public schools. Does this test count? Yes, it does count, not as a classroom grade for your current year biology class (scores come in after summer break begins), but as a way to place you in your future classes; as a way of assessing your need for extra help; as a way of helping teachers evaluate and develop their curricula and teaching; and as a way for science departments, schools, and districts to explore ways to provide the maximum opportunities to their students.

This book is written to help you achieve a passing, hopefully Advanced Proficient, score on the NJBCT. There is a lot of content included along with the critical thinking and problem solving exercises. This is because, although the NJBCT tries to focus on critical thinking and problem solving tasks, there is a lot of content and vocabulary tied into the questions, problems, and tasks. You need to be familiar with the content before you can develop a well-thought out plan to solve scientific problems.

This book is comprised of 16 chapters. Most topics are represented in multiple NJBCT areas, but their major connections are as follows:

Heredity and Reproduction
  Chapter 5: Cell Reproduction
  Chapter 6: Molecular Genetics
  Chapter 7: Mendelian Genetics
  Chapter 8: Genetic Technology
Evolution and Diversity
  Chapter 10: Classification
  Chapter 12: Evolution and Natural Selection
Interdependence
  Chapter 14: Ecology
  Chapter 15: Biomes
  Chapter 16: Human Impacts
Organization and Development
  Chapter 1: The Nature of Science
  Chapter 2: Cells
  Chapter 4: Transport in the Cell
  Chapter 11: Complex Mechanisms of Transport
  Chapter 13: Human Body Systems
Matter and Energy Transformation
  Chapter 3: Biological Chemistry
  Chapter 9: Energy

Now, it's time to get to work! As you approach the rest of this book, here are some general study suggestions: take care of yourself (eat well and get enough sleep), think positively, pay attention in class, budget your time, and organize your workload.

Good luck on the test!

CHAPTER 1

The Nature of Science

Outline for Organizing Your Notes

I. What Is Biology

II. Characteristics of Living Things
   A. Growth
   B. Development
   C. Reproduction
   D. Metabolism and the Use of Energy
   E. Organization
   F. Adaptation and Homeostasis

III. Scientific Method
   A. Observation
   B. Hypothesis
   C. Research
   D. Experimental Design
      1. Control
      2. Independent Variable
      3. Dependent Variable
      4. Placebo
      5. Double Blind Study
   E. Data
      1. Quantitative
      2. Qualitative
   F. Conclusion
      1. Theory
      2. Law

IV. Microscopy
   A. Compound Light Microscope
   B. TEM
   C. SEM

What Is Biology?

Biology is the study of all aspects of life; *bio* means "life" and *ology* means "the study of." In biology we study cells, how they work, and of what they're made. We study what it means to be alive and what characteristics we share with all other living organisms. We work to understand how energy is obtained and used by living things. In biology, we study how traits are passed from a parent cell to its offspring. We explore the body systems of living organisms and try to understand how the parts work together to function effectively. We categorize and explore the different types of living things and how they're related to each other. We determine ways that we can reduce our negative impact on our surroundings, and we examine methods and technologies to make our world a better place.

Biology is also a course that you are required to pass in high school. Some questions you may be asking yourself are: Why is Biology so important? Why do I need to learn about it? Is it as important to me as it is to my parents and teachers? As you work through this book, you will begin to understand what biology is and why it does, in fact, matter to you. As you study, keep in mind that there are many more reasons why it is important for you to understand biology than simply getting a good grade or passing a course.

Characteristics of Living Things

How do we know that something is alive? What is it that distinguishes a living organism from a nonliving thing? If you were a scientist traveling to other galaxies in search of life, would you recognize a living organism if you met one? What criteria would you use? How would you establish whether or not it was alive?

All of the above questions are easy to answer once you learn the characteristics scientists use to evaluate living organisms. Scientists agree that certain specific characteristics define and are shared by all living things. These characteristics are **growth** and **development**, **reproduction**, **metabolism** and use of **energy**, **organization**, and **adaptation/homeostasis**. As we explore each of these attributes, consider how they appear in living things. Think about how they define the difference between the living and nonliving.

Growth

All living things share the characteristic of growth. Growth is an increase in mass and size. When bacteria grow, they gain matter and get larger. When a mushroom grows, it gains matter and gets bigger. When a blade of grass, a sapling (young tree), or a puppy grows, they each gain matter and get bigger. While the amount of material gained and the increase in size varies from species to species, the fact is that all living things share the characteristic of growth.

Development

All living things undergo development. Development refers to the physical changes that happen during an organism's lifetime, with an emphasis on moving toward maturity and the ability to reproduce. One example of development in humans is puberty and the resulting potential to produce offspring. In the lifetime of a butterfly, the insect transitions from egg to larva (caterpillar) to pupa (chrysalis) to adult (butterfly). A frog's lifetime is spent developing from egg to tadpole to adult frog. As living organisms mature, they all undergo developmental changes, and although these changes may not be as dramatic as those of a human, a butterfly, or a frog, they do represent development.

Reproduction

All living things share the characteristic of reproduction. Reproduction is the production of new cells and of new individuals. In all living things, the directions for the production of new cells and offspring are carried out by the nucleic acids: deoxyribonucleic acid (DNA) and ribonucleic acid (RNA). In bacteria, reproduction involves the copying of DNA and then the asexual division of a single parent cell into two identical daughter cells, which will then grow up to be identical, or almost identical, to the original parent cell. In more complex organisms, we see both asexual reproduction (producing a new organism from a single cell parent) and sexual reproduction (producing gametes: egg and sperm).

Metabolism and Use of Energy

All living things take in nutrients, use the molecules they need, and then get rid of waste materials. Metabolism consists of two types of reactions: catabolic reactions, which break down molecules, and anabolic reactions, which build up molecules. A lion stalks and kills its prey and then eats all the parts of its body. Everything that the lion eats is chewed up and mixed with digestive chemicals that break down bones, muscle, and organs into their smallest molecules. These molecules are actively transported, or they diffuse across cell membranes to enter the lion's bloodstream. They then circulate throughout the animal's body until they get to a region where they are required. The nutrient molecules move into the cells where they are needed. They are used and any leftover waste material is transported back into the bloodstream and disposed of by the lion's excretory system. As a result of this process, the lion receives all the energy and ingredients needed to stay alive, grow, develop, reproduce, maintain homeostasis, and catch its next prey.

Organization

All living things have their bodies arranged in a specific and predictable pattern; this means that they are organized. Your dog is designed to have its eyes, ears, mouth, heart, stomach, and tail in the same places on its body as all other dogs. Your dog's body is organized. A group of daffodils will all have similar root systems, stems, leaves, and flowers all in the same places and organized in the same way. Even the simplest bacteria have organized parts: a simple cell membrane enclosing cytoplasm, a circular strand of DNA, and ribosomes.

Adaptation and Homeostasis

When a living organism maintains itself in spite of external conditions, it is adapting to its environment and is achieving homeostasis. Homeostasis occurs in humans when we maintain our body temperature in spite of external temperature changes or when our heart rate, breathing rate, and blood pressure change to accommodate both exercise and resting conditions. Emperor penguins have a system that provides heat exchange blood flow to their feet. This mechanism both keeps their feet from freezing and prevents cold blood from circulating back into the core of their bodies. Heat exchange blood flow in Emperor penguins is an example of homeostasis. Homeostasis is seen in plants when they lose their leaves and slow their metabolism in response to cold weather and when they grow toward a light source.

As we explore the science of living things, keep in mind that everything in biology connects back to these characteristics: growth and development, reproduction, metabolism and use of energy, organization, and adaptation and homeostasis.

Scientific Method

If you are looking to purchase a new computer, cell-phone, digital camera, or MP3 player, and you'd been working for a couple months to afford this purchase, would you just run out and buy the first one you saw, or would you try to consider all the possibilities, plus all the potential advantages and disadvantages of different models? Would you go online and read reviews to see how other people rate this item? How would you determine which ratings deserve your trust? What is the difference between running out and buying just any model versus researching and investigating to find a model that will best fit your needs? How is buying an electronic device similar to exploring a scientific question? Well, scientists approach their questions with similar attention and vigilance.

There is a sequence of steps in the scientific process, but it's important to remember that at every stage in a scientific investigation we can, and should, go back to revisit our question and process. We should also anticipate the steps to come. This vigilance of investigation helps scientists to ensure a thorough and rigorous exploration of the subject.

Hypothesis

Scientists approach the world with an inquisitive attitude and take in information through all of their senses, producing an **observation**. When a scientist begins to wonder about a subject that he or she has seen, heard, felt, tasted, or smelled, that scientist is questioning an observation. When a scientist has a question to be answered, he or she will explore all the possibilities he or she can think of to find an accurate and unbiased answer. Scientists develop their questions into testable statements called hypotheses. When a statement, or **hypothesis**, is thoroughly and scientifically tested, facts are established; the resulting scientific conclusion should be free of opinions, beliefs, and bias.

Research and Experimental Design

Scientists research, examine, and experiment upon their hypotheses. This process of thorough investigation allows the scientist to collect as much data as possible. There are different ways to collect and record evidence or data, including observation, research, and experimentation.

When a scientist begins to investigate a hypothesis, he or she is likely to begin by examining information already explored and published by other scientists. It's important to know what's already been done, how the hypothesis has been tested, how thoroughly the hypothesis has been tested, whether test results have been duplicated, and whether the sources of information are reputable. These questions can be answered by conducting a **literature review**. A literature review is when a scientist reads and extensively investigates the available peer-reviewed (thoroughly investigated by other scientists) published information on the topic or hypothesis. This information is usually found in academic journals or magazines. The articles in these journals include long lists of references, or citations, that can help to follow the history of how the hypothesis has been explored.

After the scientist establishes what other scientists have already done to answer the question, he or she will begin to develop an **experimental design** to test the hypothesis. The experiment must both address the hypothesis and minimize every **variable**. Variables are the factors that influence the outcome of an experiment. Scientists limit experiments to one variable. This allows them to pinpoint what influenced the results. There are several types of variables. A **control** is the constant; it is the element of the experiment that doesn't change, or that represents the original condition. An **independent variable** is the element of the experiment that is deliberately being changed. A **dependent variable** is the result of the independent variable.

In human medical research, many experiments are performed without the experimenter, or test subjects, knowing who is receiving the medicine, or product, being tested. This is called a **double blind study**. Double blind studies are often used in human medical experimentation, or clinical trials. A **placebo** is a product given to human test subjects, that looks like the medicine being tested but has no medical value. The **placebo effect** is a result of humans thinking that they're receiving a test product, or independent variable, which will have a specific effect on them. That effect then occurs even though the subjects have not been given the medicine. These elements of experimental design help to eliminate variables controlled by human choice and imagination. Scientists will do their best to minimize any extra variables, opinions, and belief systems that may influence their conclusions.

Example

Let's say you want to determine which type of painkiller works best to relieve headache pain. You might design an experiment by first establishing a large test group. You want to minimize variables, so you'll be looking for people who have similar headaches, perhaps people of the same gender or age group. You'll want to eliminate other medical factors, and you'll want to determine group size. Next you will divide the test group; you'll need a control group and a group for each variation of the independent variable (if the variable is "type of pain reliever," the choices might include aspirin, acetaminophen, and another pain reliever called a non-steroidal anti-inflammatory drug, or NSAID). As an ethical scientist, you don't want to know who is in each group and your test subjects won't know either because it is a double blind study. You can now test how quickly, or completely, each type of pain pill works.

Recording and Reporting Data

Any information gathered in an experiment is called **data**. Scientists collect two main types of data, **quantitative** and **qualitative**. Quantitative evidence includes all numerical information. Numerical and counted data can usually be collected and presented in a table or line graph format. Non-numerical, or qualitative, data includes evidence we collect using our senses. Qualitative evidence includes colors, shapes, odors, feelings, and textures and is usually presented using descriptive words and photographs.

Back to That Example

Now you've begun to collect data. Is pain qualitative or quantitative evidence? Since quantitative evidence is measurable, you want to measure the intensity and improvement of a headache. Can you think of a way to quantify pain? One way that medical researchers quantify pain is to ask the participant to rate the pain on a scale from one to ten, with ten being the worst pain ever felt. While this is a valid way of measuring pain, we do have to take into account that the scale is unique to that person. This method of data collection introduces the variable of individual perception of pain.

Conclusions

Scientists use all of this collected information to try to answer their initial question and hypothesis. If the question is answered, the hypothesis addressed, and the results appear to be significant, the scientist will usually begin writing up the work for peer review and potential publication. Often the question isn't answered, the hypothesis isn't addressed, the experiment isn't designed appropriately, too many variables or unexpected variables arise, or the results seem to be vague or insignificant. When this happens, the scientist returns to the initial hypothesis and revises each step. This may sound repetitive and perhaps boring, but it's the job of a scientist to ask good questions, and then carefully search out correct answers.

Ultimately, if a hypothesis is researched and tested repeatedly by many scientists, and the results are always substantiated, then the hypothesis may be related to other similar substantiated hypotheses and called a **theory**. Many people are confused by the scientific use of the term theory. A nonscientific theory is a guess, or speculation; a scientific theory is a group of related observations that has been so thoroughly substantiated by separate groups of scientists that the scientific community generally accepts it as an explanation of how nature works. A scientific **law** is a single rule of science; it is a statement of a specific scientific principle that describes what will happen in nature every time very specific circumstances are met. An example is the Law of Conservation of Energy, which states that the total amount of energy in an isolated system remains constant over time, meaning that it cannot gain or lose energy. The only thing that can change is the state of the energy.

The Tools of the Biologist

Scientists have many devices accessible to them for experimenting and investigating. To a biologist, the microscope is one of the most common and indispensable instruments available for exploring the living world.

Microscopy

Microscopes are one of the most important tools of the biologist. This is because biologists want to understand how living things are put together and how the component parts work together to keep an organism alive. The word microscope comes from *micro* meaning "small" and *scope* meaning "to look". The compound light microscope is the simplest and most common type of microscope used by biologists today. It was first developed around 1595, when the Jansen brothers realized that they could increase magnification by looking through two sequenced magnifying lenses. The name of this type of microscope refers to the compounding of more than one lens for magnification and the use of light to see a specimen. A compound light microscope generally provides a magnification of 100×–2,000×. This means the magnification is 100 to 2,000 times the actual size of the specimen. Calculating the magnification simply involves multiplying the magnification of the ocular lens, or eyepiece, by the objective lens (near the specimen) being used.

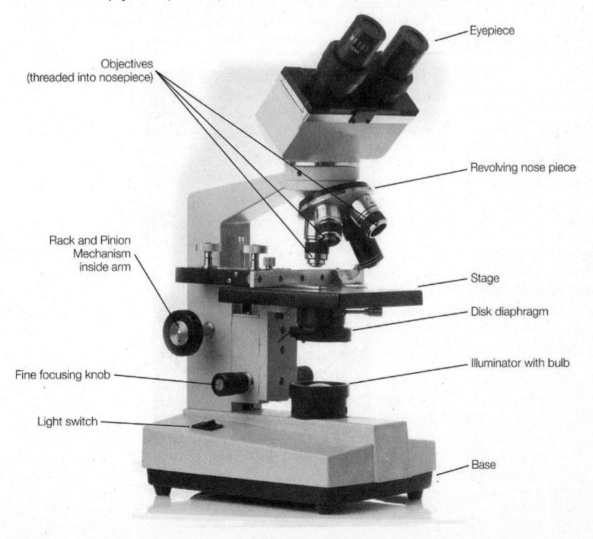

Light Microscope

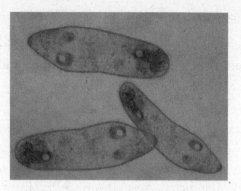

Light Microscope Image of Paramecia

Electron microscopes allow biologists to view specimens at a much stronger magnification than light microscopes. Electron microscopes use electrons, which are directed by magnetic fields, to image a specimen. The Transmission Electron Microscope (TEM) usually magnifies a specimen up to 750,000×, or even 1,000,000×. Specimen preparation for TEM imaging involves preserving, or "fixing," the sample and slicing it using a microtome, an instrument like the one that a jeweler uses to cut diamonds, and then finally attaching the transparent slice of specimen to a copper, gridded slide. Images from TEMs show us a cross section, or interior slice of the specimen. A Scanning Electron Microscope (SEM) can magnify a specimen up to about 250,000×. This microscope does not require the specimen to be sliced, and it produces a three-dimensional image of the specimen.

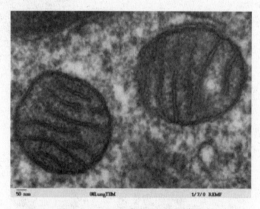

TEM image of mitochondria

SEM image of an ant head

Important Terms to Remember

Adaptation: The ability of a living thing to change some aspect of itself to accommodate its circumstances or environment.

Control: The condition in an experiment that does not change. It can be used to measure change in a dependent variable.

Data: Information collected from research or experimentation.

Development: A characteristic of living things in which their anatomy changes as they get older.

Double blind study: A human experiment in which all subjects are known only by a code name. Neither experimenter nor test subject knows who is receiving a product or who is not receiving the product (acting as a control).

Experimentation: The scientific process of controlled testing, minimizing variables, recording, and analyzing data, and the retesting to check for reproducible results.

Growth: Life process of gaining mass and size.

Homeostasis: Life process involving the adaptations made by an organism to maintain the body's balance, or equilibrium, in spite of changing external conditions. Does your heart beat faster when you exercise really hard? This happens because your cells need more oxygen than when you're sitting around, so the heart pushes out more red blood cells to provide this oxygen.

Hypothesis: An early stage of the scientific method in which a question is turned into a statement that predicts an outcome.

Law: A single fact of science; thoroughly tested, substantiated, and accepted. A scientific law tells us what will happen every time specific circumstances are encountered.

Literature review: A scientific process of investigating peer-reviewed journals to research a question or hypothesis. Peer-reviewed journals are highly respected and credible because their content has been thoroughly investigated by experts in the field.

Metabolism: Life process in which nutrients are taken in and broken down to small enough molecules to be used by the cells in a body; after nutrient molecules are used, waste products are eliminated.

Observation: The action of paying careful attention to information taken in by our senses; all information is then recorded as accurately and honestly as possible

Organization: All living things are arranged in specific ways. Living cells have their organelles arranged in a specific internal pattern. Multicellular organisms have their cells, tissues, organs, and systems located in very specific locations. Aren't you glad your fingernails are at the ends of your fingers and not at the bases?

Placebo: In a medical experiment, on humans, it can be important to treat every test subject in the same way. This means, for example, that if anyone gets a pill, then everyone should get a pill, even if some of them have no pharmaceutical ingredients. This is sometimes called a "sugar pill."

Placebo effect: When a person is given a "sugar pill," it may still have the expected effect. This happens because the human mind can be very sensitive to suggestion. If members of a test group are given a pill and told it will have a specific effect, the effect is likely to be seen even in the group that received a non-medicated pill.

Qualitative data: Scientific information gathered from sensory input: colors, shapes, odors, sounds, and textures.

Quantitative data: Scientific information gathered in numerical form, often presented in the form of tables, graphs, and charts.

Reproduction: Life process in which new cells, gametes, or organisms are formed.

Theory: An explanation of how nature works. Related hypotheses that have been tested and substantiated so many times that they are accepted as a scientific foundation. A "scientific theory" is not the same as a "conversational theory," which is a presumption or a guess.

Variable: The elements of an experiment that can change or be altered. Types of variables include control, independent, and dependent variables.

Practice Questions

Multiple-Choice

1. Your teacher gives you three living specimens to compare. One is a caterpillar, one a cocoon (or chrysalis), and one is a full-grown butterfly. You observe them and do some research. You realize that all are different stages in the life of a Monarch butterfly. Which characteristic of life are you observing?

   (A) growth
   (B) organization
   (C) development
   (D) reproduction

2. The iguana that your teacher keeps in your biology classroom is a vegetarian. It seems to have a lot of energy after it eats a big meal of kale, sweet potato, apple, and carrots. In fact, after it ate today, it was running all around its cage. Which characteristic of life does this behavior exemplify?

   (A) metabolism
   (B) organization
   (C) homeostasis
   (D) development

3. While your friend was skateboarding a week ago, he fell and scraped the palm of his hand. Now that the scab has started coming off you can see the new skin that has formed. New skin has been formed as a result of which characteristic of life?

   (A) reproduction
   (B) metabolism
   (C) homeostasis
   (D) development

4. You and your lab partner have been wondering whether you can grow a plant in the dark. Your lab partner thinks that if you give the plant the sugar it would make during photosynthesis, then it shouldn't need the sunlight anymore. Before you begin your experiment, your science teacher wants you to develop a statement about what you think will happen. What is this statement called?

   (A) hypothesis
   (B) question
   (C) theory
   (D) data

5. Your biology teacher has explained that even though we can't travel back in time to the beginning of the Earth, we do have a very good idea of the events that occurred as the Earth formed into what we know today. Our knowledge of early Earth comes from many experiments that were repeated by many scientists and validated many times. Information that is substantiated, but not considered irrefutable is called a(n) —————.

   (A) hypothesis
   (B) law
   (C) theory
   (D) proposal

6. You've designed an experiment about the way that flowers attract pollinating insects. You will be collecting information about flower color, fragrance, and the number of bees landings on each flower in your study. Your information on color and fragrance is _____ data.

   (A) insignificant
   (B) numerical
   (C) quantitative
   (D) qualitative

7. Your aunt tells you about a study in which she's been a participant. The study will determine the effectiveness of sleeping pills in a human test group. The 100 participants each take a pill at bedtime, half receive sleeping medication, and half take a pill without medication. No one, even the doctors running the study, knows which kind of pill anyone gets. What is the scientific name for the pills that don't contain medicine?

   (A) test pills
   (B) placebo
   (C) neutral medicine
   (D) double blind pills

8. You've designed an experiment to see if plant food will make a plant grow larger. You buy several similar sized bean plants. You give all of them equal sunlight and water. Half of them are given equal amounts of plant food and the other half gets no plant food. In this experiment the plant food is the _____.

   (A) control
   (B) dependent viable
   (C) independent viable
   (D) placebo

9. In the plant feeding experiment in Question #8, what is the role of the plants that were not given plant food?

   (A) control
   (B) dependent viable
   (C) independent viable
   (D) placebo

10. You've been given photographs of a fly's eye and your assignment is to try to count its facets, or lenses. You've been able to count about 750 facets! You can tell that these photographs were produced through a technology called _____.

    (A) digital photography
    (B) infrared imaging
    (C) SEM
    (D) TEM

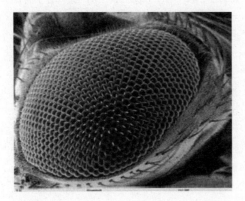

Free-Response Questions

1. Your family has decided to adopt a puppy from the local shelter. Thoroughly describe the characteristics of life that you will be able to observe as your new puppy grows up.

2. You have had a small dry area of dirt in your backyard, where nothing ever seems to grow. You wonder why this happens, and then decide to do an experiment to figure out the reason why you can't get anything to grow there. How will you approach your gardening investigation?

3. You have accepted an opportunity to participate in a marine exploration summer research project. The main job of your team in this project is to microscopically examine water and sea floor samples and to identify all living organisms in your samples. You've found a couple of specimens that no one can identify. Are they even living organisms? How would you know?

Answer Explanations

Multiple-Choice

1. **C** You're looking at a caterpillar, a chrysalis, and a full-grown butterfly; all are different stages in the life of a Monarch butterfly. If you are looking at three unique and different stages of a butterfly's lifetime, then you are observing stages of **development**. Growth would be an increase in size, or mass. If you were observing organization, you'd be looking at how the parts were put together. If it were reproduction, you would be seeing the appearance of new cells or a new organism.

2. **A** After the iguana ate a big meal, it was running all around its cage. All this energy came from the breakdown and absorption of nutrients from **metabolism**. Its energy would not come from simply being assembled in a predictable pattern, from being able to maintain itself in spite of external conditions during homeostasis, or from changes that occur in its body as it gets older.

3. **A** New skin that forms following an injury is the result of cellular **reproduction**, making new cells. Metabolism nourishes those new cells and it gives them energy, but it doesn't produce new cells. Homeostasis is the process those cells use to stay in balanced in spite of external conditions, and development is the set of changes that will occur in those cells as they age.

4. **A** When you make a statement about what you think will happen in a scientific experiment, it's called a **hypothesis**. It is a question reworded into a statement that describes a predicted outcome. If the hypothesis were to be tested repeatedly by many scientists and always substantiated or validated, it could become a scientific theory. Your statement about what you expect to happen will need to be supported by the data, or information, you collect in your experiments.

5. **C** When many related experiments are repeated by many scientists and validated many times to the point where the information is considered an accurate description of how nature works, it is called a **theory**. Earlier in the scientific process, this scientific question was developed into a testable statement called a hypothesis. Experiments are the mechanism for testing the hypothesis. Proposal is a non-scientific term for suggesting that something might happen.

6. **D** Data, including color and fragrance, are called **qualitative** data. Color, fragrance, patterns, textures, and shapes are all important observations. Numerical, or quantitative, data are different types of information and are also important.

7. **B** The scientific name for the pills that don't contain medicine is **placebo**. The terms "test pills" and "neutral medicine" are not scientific terms. Double blind refers to a type of clinical trial in which neither the scientists nor the test subjects know who is receiving which drugs, it does not refer to a type of pill.

8. **C** In this experiment, half of the subjects are given equal amounts of plant food and the other half get no plant food. The plant food is the element of the experiment that is being changed, it is the **independent variable**. The controls are the plants that receive no plant food and the dependent variable is the different growth rates of fed and non-fed plants. Placebos are only used on human subjects.

9. **A** In the plant feeding experiment in Question #8, the plants that were not given plant food acted as **controls**. Again, plant food is the independent variable, growth rate is the dependent variable, and plants don't receive placebos, they only work on humans.

10. **C** A photograph of a fly's eye that allows you to count its facets would have been produced through a technology called **SEM** (scanning electron microscopy). Digital photography and infrared imaging are not microscopic imaging technologies. Transmission electron microscopy, or TEM, would provide you with cross sections rather than a surface view of your specimen.

Fee-Response Suggestions

1. **Your family has decided to adopt a puppy from the local shelter. Thoroughly describe the characteristics of life that you will be able to observe as your new puppy grows up.**

   - **Growth:** The puppy will get bigger, gaining weight and mass.

   - **Development:** Puppy teeth will fall out and adult teeth will grow in their place; the dog's body will become more coordinated and capable of producing offspring.

   - **Reproduction:** As the puppy grows and gets larger, its cells will be reproducing through mitosis and cytokinesis (asexually) to form new cells that are identical to its parent and sister cells. Of course, you aren't likely to see this happening. If you don't have your puppy neutered, it may someday have puppies of its own, and this is an example of sexual reproduction.

   - **Organization:** You can see from day one that your puppy's body parts are organized. Its eyes, ears, toes, and teeth are arranged similarly on both the right and left sides of its body. Its parts are arranged similarly to those of its littermates and to other dogs.

- **Homeostasis:** If your puppy is playing hard, you'll notice that it breathes more heavily and its heart beats faster. When it's sleeping, or relaxing, its breathing and heart rate are slower and less noticeable. These changes allow your puppy's body to stay in balance even though its circumstances change.

- **Metabolism:** If your puppy eats its food, drinks its water, eliminates waste and seems healthy, you can be reasonably certain that its metabolism is working effectively. It is taking in nutrients, breaking them down, absorbing them into cells that need them, building cells, obtaining energy, and then, finally, getting rid of waste.

2. **You have had a small dry area of dirt in your backyard, where nothing ever seems to grow. You wonder why this happens, and then decide to do an experiment to figure out the reason why you can't get anything to grow there. How will you approach your gardening investigation?**

   - **Observation:** One section of yard where nothing grows.

   - **Hypothesis:** If plant food is added to the area where nothing grows, then some plants will start to grow there.

   - **Research:** Online, talking to people who are successful at growing plants, talking to someone that sells plants and plant products, gardening magazines.

   - **Experiment:** Divide the non-growing section in half; treat both sides the same except add plant food to one side. Take equal care of both sides.

   - **Data Collection:** Measure plant height or count plants on each side. Count or estimate plant density. Put data into table, graph, or chart.

   - **Conclusion:** Describe factual results. Do not include opinions.

3. **You have accepted an opportunity to participate in a marine exploration summer research project. The main job of your team in this project is to microscopically examine water and sea floor samples and to identify all living organisms in your samples. You've found a couple of specimens that no one can identify. Are they even living organisms? How would you know?**

   - Are they made up of organized parts?

   - Are there indications of growth, development, reproduction, metabolism, or homeostasis?

   - Do you see evidence of change or movement?

   - Are you finding more than one of any type of specimen?

- What is the magnification and type of microscope that you're using to view your specimens, if any?

- Have you done any research in peer-reviewed journals or at university websites to see if anyone else has ever seen anything similar?

CHAPTER 2

Cells

Outline for Organizing Your Notes

I. What Is a Cell?

II. History and the Cell Theory
   A. Understanding the Cell
   B. Developing the Cell Theory
   C. Levels of Organization

III. Types of Cells
   A. Prokaryotes
   B. Eukaryotes

IV. All Cells Need Nourishment
   A. Autotrophs
   B. Heterotrophs

V. Organelles
   A. Holding It All Together
     1. Cytoplasm
     2. Cell Membrane
     3. Cell Wall

   B. Cellular Control and Protein Synthesis
     1. Nucleus
     2. Nuclear Membrane
     3. Nucleolus
     4. Ribosome
   C. Energy Production and Storage
     1. Mitochondria
     2. Chloroplast
   D. Chemical Processing and Transport
     1. Endoplasmic Reticulum
     2. Golgi Apparatus
   E. Storage and Waste Disposal
     1. Lysosome
     2. Vacuole
     3. Peroxisome
   F. Structure, Support, and Movement
     1. Microtubule
     2. Microfilament
     3. Centriole
     4. Cilia
     5. Flagella

What Is a Cell?

When we look at another living organism, we may be looking at billions of microscopic cells working together. Although it may not seem important for you to understand the collection of living cells you're looking at, it would be, if you were a medical person. You would want to know how the cells work, or don't work, together in order to help them stay healthy. If you were a scientist, you would want to understand how the parts of the cell influence the workings of the whole cell and then, of the whole body. As a student, you want to have a general understanding of how living things function and stay alive. As a member of society, you want to recognize how to best care for other living things and how to maintain the environment so that living things can thrive.

All living things are either cells or are made of cells. So, what does it mean to be a cell? What is it about cells that make them special and unique? How are cells different from each other and how are they alike? Are there characteristics of life that are more important than others? How have we come to our current understanding of life and how might that understanding change in the future?

History and the Cell Theory

Until the earliest microscopes were developed in the late 1500s, there was no tangible evidence to suggest the existence of cells, or of structures, or organisms, too small to see with the naked eye. Several scientific figures are known for their contributions to our early understanding of the microscopic world. Robert Hooke (1635–1703) used a compound light microscope to view a piece of oak tree bark, called cork. He compared the patterns of many, identical, small rectangular structures that made up the cork, and thought that they looked like the rooms that monks lived in at a monastery; these rooms were called "cells." Hooke named the structures he'd discovered in the cork, "cells." In 1665, Hooke published many of his drawings of images he had seen through his microscopes in a book called *Micrographia*.

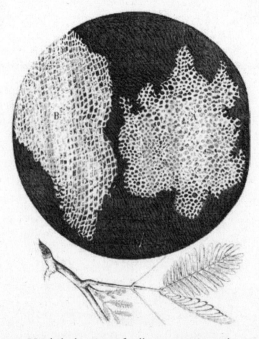

Hooke's drawing of cell structure in cork

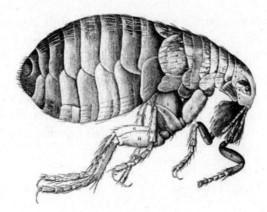

Hooke's drawing of a flea

Another important figure in the revolution of microscopy was "the Father of Microbiology," Anton Van Leeuwenhoek (1632–1723). Van Leeuwenhoek developed and improved compound light microscopes. Using them, he became the first person to examine and describe single-celled microorganisms, which he called "animalcules." Van Leeuwenhoek was the first person to observe and record microscopic images of bacteria, spermatozoa, muscle fiber, and capillary blood flow.

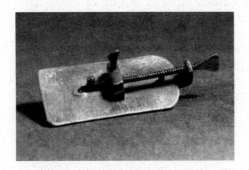

Van Leeuwenhoek's microscope

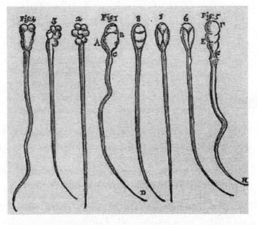

Van Leeuwenhoek's drawings of spermatozoa

Another scientist who was important in the development of our understanding of cells was a physician named Francesco Redi (1626–1697). Redi conducted a series of experiments that challenged the then-popular belief in abiogenesis, or "spontaneous generation." Until Redi's work in the 1660s, it was commonly believed that flies and maggots arose from rotting meat. Redi tested this idea by placing meat into six jars; three jars were left open to the air and the other three were covered with fine gauze so that only air could enter. Several days later, maggots appeared on the uncovered pieces of meat but not on the gauze covered meat. Redi kept the maggots and allowed them to develop into flies. Redi also placed living and dead flies into sealed jars with meat, recognizing that maggots only developed in the jars containing living flies.

Developing the Cell Theory

Between the mid-1600s and the mid-1800s, scientists explored the microscopic world but were somewhat unsure of the relationships between cells. Scientists did not understand the interactions between different cells, between cells and tissues, between cells and organisms, and they were unsure of the predictability of these relationships. Ultimately, the Cell Theory was developed and became the accepted explanation for the role of cells and how they function.

The Cell Theory states that:

- All living things are composed of one or more cells. (Schleiden and Schwann)

- Cells are the basic units of structure and function in living things. (Schleiden and Schwann)

- New cells are produced by existing cells. (Virchow)

Three scientists have been credited with developing the cell theory. Matthias Schleiden (1804–1881) was a botanist who, in 1838, proposed that cells are the basic unit of structure and function in all plants. Theodor Schwann (1810–1882) was a zoologist who, in 1839, established that cells are the basic unit of structure and function in all animals. Rudolf Virchow (1821–1902) was a physician and pathologist who, in 1858, concluded that all cells come from other living cells. Together, these scientists formulated the foundation for our current understanding of cellular structure and function, the Cell Theory.

Levels of Organization

Based on the Cell Theory, we've established that cells are the smallest units of structure and function in living things. Some organisms are **unicellular**, or single-celled, and some are **multicellular**. A unicellular organism has the ability, within its one-celled body, to perform all the functions necessary to maintain life. A multicellular organism has cells that perform specific functions within its body, so all cells do not do the same jobs. In a multicellular organism, cells that perform similar functions are grouped together and called tissues. Tissues that are grouped together to perform a function are called organs. Organs that are grouped together and have similar structure and function are called organ systems. When organ systems work together, they form a complex organism.

Types of Cells

Prokaryotes

The simplest type of cell is the **prokaryote**. All bacteria are prokaryotes and all prokaryotes are bacteria. Prokaryotes are all single-celled organisms that do not contain complex membranes. This lack of complex membranes means that prokaryotes do not have any internal barrier surrounding their DNA. Prokaryotes do not have a nucleus. Bacterial DNA exists in the cytoplasm. Other structures found in prokaryotes include cytoplasm, ribosomes, a simple membrane to contain the cytoplasm, a cell wall, and sometimes, an external wall or capsule.

Bacteria live in all environments on Earth and belong to the kingdom(s) with the most members on Earth. There is tremendous variety among the species of bacteria. Bacteria, or prokaryotes, are always single-celled and can be either **heterotrophs** or **autotrophs**. Prokaryotes can be stationary (nonmoving) or mobile. They may be harmful to other living things (including humans) but most are helpful.

Eukaryotes

All cells that are not prokaryotes are **eukaryotes**. If a cell is a eukaryote it has complex membranes surrounding its internal structures, or organelles. Specifically, eukaryotic cells have a membrane-bound nucleus that contains most of their DNA. Eukaryotes are more complex than prokaryotes and make up the Kingdoms Protista, Fungi, Plantae, and Animalia. Eukaryotic organisms may be unicellular or multicellular. If a living thing is unicellular, or single-celled, it must contain all the parts needed to keep itself alive. A multicellular organism is likely to have more specialized cells that perform unique functions within the organism.

All Cells Need Nourishment

Autotrophs

Autotrophic cells make their own food. Many autotrophs make their food by conducting photosynthesis. During photosynthesis, cells take in energy from sunlight and use that energy to rearrange carbon dioxide (CO_2) and water (H_2O) into sugar ($C_6H_{12}O_6$) and oxygen (O_2). Organisms that are photosynthetic autotrophs include all plants, some protists, and some bacteria. Some autotrophs are chemosynthetic instead of photosynthetic. Organisms that perform chemosynthesis use methane or inorganic molecules like hydrogen gas or hydrogen sulfide as a source of energy to create nutrients. When a living thing makes the nourishment it needs, it is called a producer.

Heterotrophs

If a cell doesn't make its own food, then it must get nourishment from something that does. Organisms that do not produce their own food are called heterotrophs. Heterotrophs take in the carbohydrates (or sugars) and oxygen produced by autotrophs and use them to get energy to keep their cells functioning. This process is called cellular respiration. A heterotroph can be a consumer, a scavenger, or a decomposer; a consumer cannot internally manufacture its own food and therefore, obtains its nutrients from eating another organism (either an autotroph or heterotroph). A scavenger begins the process of decomposition by eating dead plants or animals, and a decomposer gets its nutrients by breaking down already decaying material.

Organelles

Within eukaryotic cells are **organelles**; structures that work for the cell in the same way that organs work within the human body. These organelles are organized inside the cell, allowing each cell to perform its specialized functions. The job of the cell determines which organelles will be more, or less, densely assembled in that cell.

An Analogy

Cells are like communities. They are made up of many parts, each part performing a specific function or task. As we move through each cellular organelle, try to figure out what part of a community the organelle resembles.

Holding It All Together

The semi-liquid material contained by the cell membrane is called **cytoplasm**. This soft, transparent, gelatin-like material is the site of many of the cell's continuous chemical reactions. Cytoplasm provides suspension for organelles inside the cell; it maintains the cell's shape and consistency. The cell and most of its organelles are each surrounded by a membrane that selects what will enter and leave the cell, or organelle. The same type of **cell membrane** is found surrounding both the cytoplasm and the organelles. The cell membrane is made of a phospholipid bilayer with embedded proteins and cholesterol. Phospholipids are molecules that interact well with water at one side and interact well with lipids, or fats, on the other side. Embedded proteins provide pathways for large molecules to move into or out of the cell and cholesterol molecules hold the two phospholipid layers in a relatively stable arrangement. Some cells, including all plant cells, have a **cell wall** on the outside of their cell membrane. The cell wall provides strength and structure to the cell while allowing some materials to filter into the cell. Cell walls are part of the reason that a tree can grow to be 100 feet tall without collapsing down on itself. Some bacteria, protists, and fungi have them, but animal cells do not have cell walls.

The Cell Analogy

If the cell is like a community, then the cell membrane would be like the city limits, or like the borderline around the town. The cell wall would be like the relatively impenetrable super highway "loop" around the city, and the cytoplasm would include the land that the town is built upon.

Cellular Control and Protein Synthesis

The cell is directed by information found in its **nucleus**. This information is carried in molecules of deoxyribonucleic acid (DNA). The nucleus is surrounded by a porous **nuclear membrane** composed of, and interchangeable with, the same type of phopholipid bilayer as the cell membrane. Semi-liquid material contained by the nuclear membrane is called nucleoplasm. Nucleoplasm suspends materials in the nucleus, gives the nucleus structure, and provides a medium for chemical reactions that occur in the nucleus. This soft, gelatin-like substance holds DNA and the nucleolus in place and provides for the movement of molecules into the cytoplasm. Suspended in the cell's nucleoplasm is a non-membrane bound structure called the **nucleolus**, the site of **ribosome** production. Ribosomes are made of ribonucleic acid (rRNA) and proteins. After being assembled, the ribosome leaves the nucleolus and the nucleus and is the cytoplasmic site of protein synthesis, or manufacture.

> **Continuing with Our Cell Analogy**
>
> The nucleus is the control center, probably the government headquarters or the city hall building. Nucleoplasm is represented by the offices inside of city hall, the location where city planning, building plans, and laws are finalized. City supervisors and management are represented by the nucleolus, setting up the plans and foundations for city activities. Ribosomes are like an assembly or manufacturing plant that makes parts to help the city run smoothly.

Energy Production and Storage

Cells need energy for almost every task they perform: work, taking and using nutrients, extracting and getting rid of wastes, growth, development, transport, movement, healing, homeostasis, and building new molecules or cells. The **mitochondria** in a cell manufacture and store energy in the bonds of a molecule called adenosine triphosphate (ATP). Mitochondria have a double membrane that projects inward into the organelle providing a lot of surface area for ATP production and storage. These projections are called christae.

Autotrophs have a unique organelle, the **chloroplast**, used to conduct photosynthesis. Chloroplasts have internal structures that hold chlorophyll, a pigment containing compound, which captures light energy from the sun. The light energy is transformed to chemical energy as CO_2 and H_2O are split apart and rearranged into sugar and O_2.

> **The Cell Analogy**
>
> All cities need energy. It is provided to the city by power plants, transforming stations, and electric companies, just as it's provided to the cell by mitochondria. Cities use food manufacturing plants to convert the raw materials of food into usable food sources, just as chloroplasts convert water and carbon dioxide into usable sugar and oxygen.

Chemical Processing and Transport

The **endoplasmic reticulum** is a tubular network that carries cellular materials wherever they're needed in the cell. There are two types of endoplasmic reticulum: rough endoplasmic reticulum, RER, and smooth endoplasmic reticulum, SER. Rough ER looks bumpy, or rough, on the outside because ribosomes are embedded in its surface. These ribosomes produce proteins that are then moved around the cell by the RER. Smooth ER is primarily involved in metabolism and transport of carbohydrates, and in the manufacture and transport of lipids and steroids.

Another system of stacked and flattened, tube-like storage structures in the cell is the **Golgi apparatus**. The Golgi apparatus, or Golgi complex, receives vesicles of the molecules from the ER from inside the cell, and essentially sorts, labels, refines, and packages them into secretory vesicles, to be delivered where they're needed. Secretory vesicles are sacs produced when the Golgi apparatus is ready to send out a chemical delivery by pinching off a section containing that molecule. These vesicles are surrounded by membranes that have come from the Golgi apparatus but are just like all the other membranes in the cell. This allows them to insert themselves into, or attach to, other organelles or to the cell membrane. So, if the secretory vesicle contents are supposed to be disposed to the cells exterior, it will simply join with the existing cell membrane and discard its contents to the outside.

The Cell Analogy

Moving around inside a cell is like moving around a city. You can't always just start moving and end up where you need to be; sometimes you need vehicles, perhaps rail lines, to transport you or sidewalks and streets to help guide you. Endoplasmic reticulum is like mass transit. The trains and subways are like the tubes of ER, transporting materials to areas of the cell where they are needed. The Golgi apparatus is like the post office, sorting, packaging, and sending materials off to its final destination. Secretory vesicles are like post office trucks, loaded up with materials and destined for their delivery.

Storage and Waste Disposal

Cellular metabolism produces nutrient molecules. Some of the particles are stored and some are eliminated as waste. The **lysosome** is a type of secretory vesicle that pinches off from the Golgi complex; it temporarily stores and digests nutrients, and it digests and eliminates waste for the cell. It contains mild acids and has digestive enzymes to break down particles of food, unusable or old cell parts, and any other materials that can be metabolized and used by the cell or digested, ultimately eliminating any waste products. **Vacuoles** are compartmentalized, water and enzyme filled organelles that are more prominent in plant than animal cells. They store some molecules and eliminate waste products. Vacuoles are important to plant cells because they help them hold their shapes by providing internal pressure against the plant cell wall. Have you ever given water to a wilting plant? You can almost immediately see the plant filling out again as the water refills its emptied vacuoles.

Peroxisomes are a group of organelles derived from the endoplasmic reticulum that contain enzymes that eliminate potentially lethal peroxide from cells. Peroxisomes are also involved in breaking down fatty acids and synthesis of bile acids (helps in digesting fats), myelin (an insulating coat around our nerve cells), and cholesterol (helps keep the cell membrane stable).

The Cell Analogy

In every city, there are locations where food is stored and kept available and places where waste is processed. Lysosomes serve a double duty in the cell; they act both as restaurants and as recycling and sewage treatment plants. Peroxisomes serve the cell as hazmat disposal facilities. Vacuoles perform the function of a reservoir and water treatment plant.

Structure, Support, and Movement

One of the characteristics of living things is organization. In order to maintain its shape and organization, the cell needs to maintain a supportive structural system. Many cells also need to be mobile, or capable of movement, in order to get to their nutrients or to avoid becoming someone else's food. The organelles of the internal and external cytoskeleton provide the structures to allow these functions. **Microtubules**, intermediate fibers, and **microfilaments** are organelles of the internal cytoskeleton. They provide a scaffold within the cell to hold structures in place, to move other structures to where they belong, and they serve as a path for some organelles to travel along inside the cell. Microtubules are thick, hollow tubes that provide support and shape to the cell and form tracks for other organelles to move around in the cytoplasm. Microtubules and microfilaments are held in place by intermediate fibers. Microfilaments are solid, thread-like rods that also support the cell, aid in muscle contraction, and maintain its shape.

The **centriole** is also an internal structure made up of microtubules, found only in animal cells. Two centrioles direct cellular reproduction by aligning the spindle fibers during mitosis and meiosis. External cytoskeletal structures are the **cilia** and **flagella**, both made up of microtubules and used for movement; either locomotion (movement) or the action of sweeping something (usually food) past, or into, the cell. Cilia are shorter, hair-like projections found in large numbers covering the entire exterior of some cells. Flagella are longer, hair-like structures that tend to be found in smaller numbers at some specific location on a cell.

Another Cell Analogy

The cytoskeleton is similar to a transportation system because it directs movement inside the cell as it provides structure and support to the cell. Microtubules are like highways and microfilaments are like streets, both helping organelles and materials to move around inside the cell. Centrioles direct traffic during mitosis acting like traffic lights and signs. Cilia are like access roads from out of town into industrial areas. Flagella are similar to railroad and subway tracks.

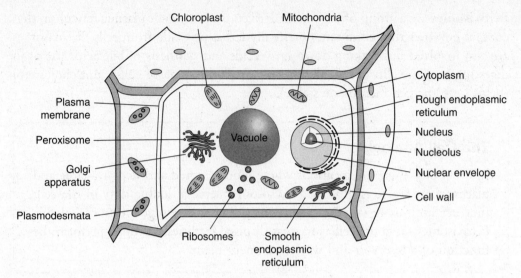

Anatomy of the Plant Cell

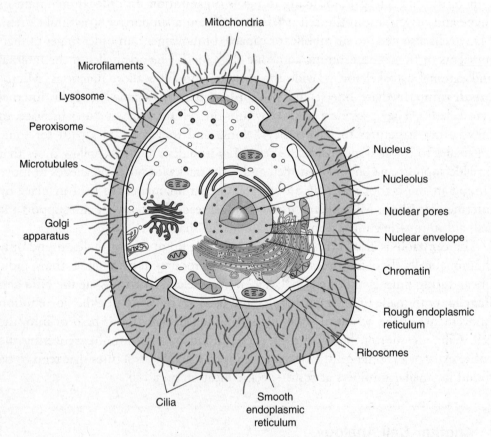

Anatomy of the Animal Cell

Completing the Cell Analogy

Organelles are important to the cell in the way that land, roads, buildings, and people are important to the community. Likewise, specific molecules are important in making cellular activities happen and specific people are important in making the community function effectively. DNA, which contains all the instructions for the cell, is similar to the mayor and city council, directing activities, establishing rules, deciding who will be responsible for the activities, and when they should happen. RNA, which takes the directions from DNA, and leaves the nucleus to direct the assembly of proteins, is like the city employees, implementing the mayor's directions. Proteins are similar to the citizens, people living in town who spend their time doing their various jobs making the community functional.

Cells in Different Kingdoms

Bacterial Cells

All bacteria are prokaryotes. They do not have a nucleus or complex membranes. They have a simple cell membrane, with DNA suspended in their cytoplasm, and ribosomes to make proteins. They may, or may not, have a cell wall. They may be stationary or mobile. They may be autotrophs or heterotrophs.

Protist Cells

All protists are eukaryotes. They have a nucleus and organelles. Protists can be animal-like: heterotrophic and mobile. They can be plant-like: autotrophic, photosynthetic, and stationary. Protists can also be fungus-like: stationary, heterotrophic decomposers.

Fungal Cells

All fungi are eukaryotes. They have a nucleus and organelles. Fungi are stationary, heterotrophic decomposers.

Plant Cells

All plants are eukaryotes. They have a nucleus and organelles. They are autotrophic producers. Plants are not mobile; they are stationary. Plants have chloroplasts, cell walls, and large central vacuoles.

Animal Cells

All animals are eukaryotes. They have a nucleus and organelles. They are heterotrophic consumers. All animals are mobile even if it's just for a part of their lives. For example, sponges swim while immature, and then plant themselves and become immobile. Animals have centrioles. Animal cells do not contain chloroplasts, cell walls, or large central vacuoles.

Historical Figures in Cell Biology

Robert Hooke (1635–1703) Used an early compound microscope to identify and name the "cell."

Anton Von Leeuwenhoek (1632–1723) Developed and improved compound light microscopes.

Francesco Redi (1626–1697) Supported idea of "biogenesis" and refuted the idea of spontaneous generation.

Matthias Schleiden (1804–1881) Contributed to the Cell Theory by establishing that cells are the basic unit of structure and function in plants.

Theodor Schwann (1810–1882) Contributed to the Cell Theory by establishing that cells are the basic unit of structure and function in animals.

Rudolf Virchow (1821–1902) Contributed to the Cell Theory by establishing that all cells come from other living cells.

Important Terms to Remember

Autotroph: An organism that makes its own food through photosynthesis or chemosynthesis: a producer.

Cell membrane: Structure surrounding cell's cytoplasm, also called plasma membrane. It selects what will enter and exit the cell.

Cell wall: Tough material that surrounds the cell membrane in plants, some fungi, and some protists; it gives strength and support to these cells.

Centriole: A cytoskeletal structure found in animal cells. It helps to arrange chromosomes during mitosis and meiosis so that each new cell gets the correct chromosomes.

Chloroplast: Organelle in photosynthetic cells that traps energy from sunlight; contains structures that convert light energy to chemical energy.

Cilia: External cytoskeletal structures that look like short hairs that cover the cell's surface. They help the cell to move itself or to push material around it.

Cytoplasm: Soft, transparent, gelatin-like material that makes up the substance of the cell. It suspends the organelles and provides a medium for the cell's chemical reactions.

Endoplasmic reticulum: Tubular channels that provide a transportation network in the cytoplasm and sites for some chemical reactions. They can have a smooth outer surface (SER), or they can have ribosomes embedded in the outer surface making it look bumpy, or rough (RER).

Eukaryote: Cells with complex membranes; they have a nucleus containing DNA and belong to the Kingdoms Protista, Fungi, Plantae, and Animalia.

Flagella: External cytoskeletal structures; look like long hairs, usually found in localized areas of cell; used for locomotion, or movement.

Golgi apparatus: Sac-like structures that look like stacked pancakes in the cell; they take molecules as needed from inside the cell and sort, label, refine, and package them into secretory vesicles for distribution inside and dispersal outside the cell.

Heterotroph: An organism that does not make its own food; a consumer, it has to eat other living things to obtain nourishment.

Lysosome: An organelle that temporarily stores and digests nutrients and then eliminates waste from the cell.

Microfilament: Internal cytoskeletal structures. These solid, thread-like rods support the cell, aid in muscle contraction, and help to maintain its shape.

Microtubule: Thick, hollow tubes that provide support and shape to the cell and form tracks for other organelles to move around within the cytoplasm.

Mitochondria: These organelles convert the energy from sugar into ATP and store energy in the bonds of a molecule called ATP.

Multicellular: An organism made up of multiple cells that each perform specific functions within its body, so that all cells do not perform the same jobs.

Nuclear membrane: A porous structure that contains the nuclear contents. It is composed of the same materials and is interchangeable with the cell membrane.

Nucleolus: This organelle found inside the nucleus is the site of ribosome production.

Nucleus: Contains all the genetic material in the cell; this organelle directs all cellular activity.

Organelle: A unique structure within a cell that performs a specific function.

Peroxisome: Organelles that contain enzymes that eliminate peroxide from cells, help to break down fatty acids, and synthesize bile acids, myelin, and cholesterol.

Prokaryote: Bacteria, simple organisms with very simple external membranes. They do not contain a nucleus and so their DNA is suspended in the cytoplasm.

Ribosome: Tiny organelles that assemble amino acids into proteins according to directions from mRNA.

Unicellular: Organisms made of only one cell. This type of organism must multi-task because it performs all life functions in its one cell.

Vacuole: Compartmentalized, water-enzyme filled organelles that are more prominent in plant than animal cells. They store some molecules and eliminate waste products. They help plant cells hold their shapes by providing internal (turgor) pressure against the plant cell wall.

Practice Questions

Multiple-Choice

1. The Cell Theory explains and defines the role of cells in the natural world. It does not tell us that

   (A) cells are the basic unit of structure in all living things.
   (B) cells are made of organelles.
   (C) cells are the basic unit of function in all living things.
   (D) all cells come from other living cells.

2. You've been given a set of electron micrographs of different types of cells. The image you're looking at now shows what looks like stacked tubes speckled with black dots inside the cell. These organelles are called _____.

   (A) lysosomes
   (B) mitochondria
   (C) rough endoplasmic reticulum
   (D) nucleus

3. Muscle cells do a lot of work, so they use large amounts of energy. Which organelle would you expect to see many of in muscle cells?

   (A) nucleolus
   (B) mitochondria
   (C) Golgi apparatus
   (D) endoplasmic reticulum

4. A genetic disorder called Tay-Sachs is fatal to its victims, usually in the first few years of life. It destroys brain tissue in its victims because their cells are unable to dispose of metabolic waste. Tay-Sachs is the result of malfunction in which organelle?

   (A) lysosome
   (B) ribosome
   (C) cilia
   (D) rough endoplasmic reticulum

5. The cells on your microscope slide appear to have a cell wall and a large central vacuole. What type of cell do you think you are looking at and what type of organelle would you look for to confirm?

   (A) bacteria; nucleus
   (B) protist; mitochondria
   (C) fungus; chloroplast
   (D) plant; chloroplast

6. You were walking your dog this morning and you tripped, scraping your knee and elbow. You've learned in biology class that the main structural components of your skin are proteins. Which organelles will be working extra hard as you heal?

   (A) cell wall
   (B) ribosomes
   (C) Golgi apparatus
   (D) vacuoles

7. Every living thing shares certain characteristics with all other living things. Which of the following is not a characteristic shared by all living things?

   (A) metabolism
   (B) movement
   (C) organization
   (D) homeostasis

8. You're viewing your set of electron micrographs of different types of cells. The image you're looking at now shows what look like stacked pancakes with little pouches pinching off along the edges. These organelles are called _____.

   (A) Golgi complex
   (B) mitochondria
   (C) ribosomes
   (D) nucleus

9. You're looking at that set of electron micrographs. The image you're looking at now shows what looks like stacked tubules covered with little dots. You also see the little dots loose in the cytoplasm; many are clustered near the nucleus. The organelles that look like dots are called _____.

   (A) endoplasmic reticulum
   (B) mitochondria
   (C) ribosomes
   (D) chloroplasts

10. You're still looking at the set of electron micrographs. The image you're looking at now shows what looks like large rounded dark spots inside many of the nuclei. These structures are called _____.

    (A) centrioles
    (B) nucleoli
    (C) mitochondria
    (D) chloroplasts

Free-Response

1. You have been working in a research lab in a university. A colleague has returned from a trip into a rainforest in South America and wants you to investigate some water samples she brought back. You find unfamiliar cells in the water. What are you going to look for to try to identify what kinds of cells they are?

2. Adrenoleukodystrophy (ALD) is a sex-linked genetic disorder that results from accumulation of specific types of fats in a child's brain. These fats are not broken down and disposed of and they build up as a result. Untreated, this causes severe brain damage and ultimately death. Which types of organelle appear to be malfunctioning? Explain your answer.

3. You are working for a laboratory testing and consulting firm. You've been given a large chunk of rock-like material that was taken from a cave more than three miles deep. Your job is to figure out if there is any living material in this sample. What might you do?

Answer Explanations

Multiple-Choice

1. **B** The Cell Theory does not tell us that cells are made of organelles. It does tell us that cells are the basic unit of structure and function in all living things and that all cells come from other living cells.

2. **C** The tubes are endoplasmic reticulum, and they are dotted with ribosomes, making them RER. Lysosomes would look like fluid-filled pouches. Mitochondria have a double membrane that projects inward. The nucleus is more spherical than tubular.

3. **B** You'd see large numbers of mitochondria. The nucleolus, Golgi apparatus, and ER would not be involved in energy availability, production, or storage.

4. **A** Tay-Sachs is a "Lysosomal Storage Disease" and happens when the child's lysosomes malfunction. If ribosomes or RER malfunction, protein production and transport will be affected. Cilia that aren't working will alter the cells ability to either move or to push food into a position where it can be consumed.

5. **D** With a cell wall and a large central vacuole, you'd want to look for a chloroplast to confirm that this is a plant. Bacteria will not have nuclei or vacuoles; a plant-like protist may have a cell wall and a vacuole but if you see mitochondria, it won't establish the type of organism. Fungi don't have chloroplasts because they are not autotrophic.

6. **B** Ribosomes, which assemble proteins, will be working hard to provide you with the materials to heal. Because you belong to the animal kingdom, you don't have cell walls. Golgi bodies and vacuoles don't help you with producing proteins.

7. **B** Many living things move, but not all of them; trees and mushrooms don't move. All living things share the characteristics of metabolism, organization, and homeostasis.

8. **A** You are likely to be looking at the Golgi complex with little vesicles, or sacs, pinching off to be distributed around the cell where they're needed. Mitochondria, ribosomes, and the nucleus do not look like stacked pancakes nor do they have little pouches pinching off from them.

9. **C** Ribosomes are organelles that look like tiny dots, either loose in the cytoplasm or attached to rough endoplasmic reticulum. The endoplasmic reticulum looks like stacked tubules and the mitochondria, and chloroplasts are bean-shaped with internal membranes for energy processing and storage.

10. **B** Large, rounded dark spots inside the cell's nucleus are nucleoli, structures that assemble ribosomes. Mitochondria, centrioles, and chloroplasts all exist in the cell's cytoplasm; they are not found in the nucleus.

Free-Response Suggestions

1. **You have been working in a research lab in a university. A colleague has returned from a trip into a rainforest in South America and wants you to investigate some water samples she brought back. You find unfamiliar cells in the water. What are you going to look for to try to identify what kinds of cells they are?**

   You'll need to establish whether the cells are prokaryotic or eukaryotic by determining whether the cells have nuclei. You could also consider that bacterial cells are smaller, and have fewer internal structures than eukaryotic cells. If the cells are eukaryotic, they're likely to be protists because fungi, plants, and animals are multicellular. You'll also want to establish whether these cells are autotrophs or heterotrophs; you may consider the cell's color or the presence of chloroplasts in a eukaryote. Plant cells will contain chloroplasts and cell walls, while animal cells will not have them.

2. **Adrenoleukodystrophy (ALD) is a sex-linked genetic disorder that results from accumulation of specific types of fats in a child's brain. These fats are not broken down and disposed of and they build up as a result. Untreated, this causes severe brain damage and ultimately death. Which types of organelle appear to be malfunctioning? Explain your answer.**

The information that you have indicates that ALD appears to be a result of malfunctioning lysosomes. This seems to be the case because fats are not being metabolized and wastes are not being disposed. The organelle whose job it is to break down fats and dispose of their waste products is the lysosome.

3. **You are working for a laboratory testing and consulting firm. You've been given a large chunk of rock-like material that was taken from a cave more than three miles deep. Your job is to figure out if there is any living material in this sample. What might you do?**

You might want to try to mimic, as closely as you can, the conditions from which the sample was taken. You'll need to monitor the sample for evidence of growth, development, metabolism, reproduction, organization, and homeostasis. You may want to take small samples of this specimen to look for the presence of cells.

CHAPTER 3

Biological Chemistry

Outline for Organizing Your Notes

I. The Parts List
  A. Atoms
  B. Molecules
  C. Mixtures
  D. Solutions
  E. Bonds

II. What About Water
  A. Functions of Water
  B. pH

III. Molecules of Life
  A. Building Blocks
  B. Carbohydrates
  C. Lipids
  D. Proteins
  E. Nucleic Acids
  F. Polymers
  G. Monomers

This is biology! Why are we talking about chemistry? Well, have you ever had a big question or problem that seemed insurmountable, but then you realized you could solve it if you broke it down to its smallest component parts? When we discuss biology, we're talking about organisms whose smallest component parts are mostly carbon, hydrogen, oxygen, nitrogen, sulfur, and phosphorus. If we want to understand how living things work, or how they're put together, we need to know about these elements and how they interact with each other and with other elements.

Basic Biochemistry

In order to understand how living things are put together and how they function, it's important to understand the basic chemical structures that are common to life. Mostly, we have to understand how carbon (C) interacts with other elements in living things. These are the foundations for two branches of science called **biochemistry** and organic chemistry.

A Parts List

Atoms

The smallest functional units of matter are called **atoms**. When a substance consists of only one type of atom it is called an **element**. Elements are listed on the periodic table in a way that allows us to determine how many protons and electrons they contain, to see their relationships to each other, and to make comparisons between them. Atoms are made up of positively-charged protons, negatively-charged electrons, and neutral neutrons. The protons and neutrons exist inside the nucleus, or core, of the atom, while the electrons move throughout electron levels outside the nucleus. A stable atom has a full outermost electron **orbital**. The electron levels, or orbitals, can each contain a maximum number of electrons; the first level can only accommodate a maximum of two electrons, the second level reaches its maximum at eight electrons, and the third level takes a maximum of eighteen electrons but can be stable with eight. When an atom is unstable because its outermost orbital isn't full, it is likely to bond with other atoms to form a molecule.

Molecules

When an element mixes together with other elements, they can form a compound or a **molecule**. Salt is a compound made of sodium (Na) and chloride (Cl), and sugar is a molecule made of carbon (C), hydrogen (H), and oxygen (O). Often you'll see numbers in a description of a molecule or a formula. If the number is in the form of a subscript following the atomic symbol, it tells you how many atoms of that element are present in the molecule. If the number precedes the molecular formula, then it's telling you how many molecules are present. So, in one molecule of carbon dioxide (CO_2) you'll find one carbon atom and two oxygen atoms. If you had two molecules of carbon dioxide, they would be represented by the chemical formula $2CO_2$.

Mixtures

When molecules are combined together, but do not change or share properties, they become a **mixture**. An example of a mixture is mixing dry sugar and sand together. They each retain their separate and original characteristics, or properties. If you were willing to work hard enough, you could separate the two from each other.

Solutions

When some compounds and molecules are mixed together they may join together chemically to become a **solution**. In a solution, one (the solute) is dissolved into the other (the solvent). An example of a solution is what happens when you stir a packet of powdered iced tea mix into water. Both the water and the tea are changed into liquid iced tea. The two types of molecules can't be returned to their original states.

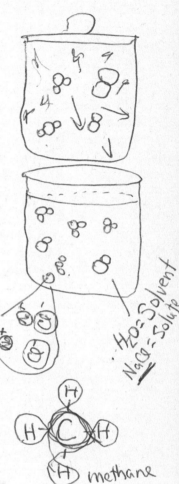

H₂O=Solvent
NaCl=Solute

Bonds

The attraction between two atoms in a molecule is called a bond. There are different kinds of bonds that form in different ways and offer different levels of strength. These bonds include the **covalent bond** in which electrons are shared between two, or more, atoms. Covalent bonds are found in the bonds of water and glucose. In an **ionic bond**, electrons are given by one element and taken by another; this happens in salts. **Peptide bonds** hold together amino acids in order to form proteins. **Hydrogen bonds** are found in many molecules, one important function of the hydrogen bond is to hold the nitrogen base groups of DNA together.

methane

What About Water

All cells need water to survive. Most cells are made up of approximately 70 percent water. In multicellular organisms, the fluid outside cells (intercellular fluid) and the fluid inside the cells (intracellular fluid) are primarily composed of water. Water serves as a solvent, giving nutrients a medium in which to dissolve and allowing them to cross through cell membranes. Water helps transport nutrients around the body to be delivered to cells that need them. Water transports waste materials out of cells and out of the bodies of multicellular organisms. Water acts as a cushion and shock absorber for cells and body parts that might be damaged by strong impact. Water helps bodies to maintain homeostasis by regulating their temperature. Water pressure helps cells to maintain their shape and internal structure.

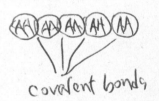

covalent bonds

pH

Healthy organisms maintain a very controlled acid-base balance in their bodies. Acids and bases are measured on a **pH scale**. The pH scale runs from 0 to 14, with 0 representing the strongest acid, 7 representing pure, neutral water, and 14 representing the strongest base. Most cellular processes occur at a pH between 6 and 8. Acids and bases are formed when water (H_2O) splits apart into two types of charged particles called ions, the hydrogen ion ($H+$) produces acids and the hydroxide ion

H^+ urine is acidic!

(OH–) makes bases. Severe chemical burns can occur at both ends of the pH scale, strong acids of 0 and strong bases at 14 will both cause burns.

Concentration of Hydrogen ions compared to distilled water		Examples of solutions at this pH
10,000,000	pH = 0	Battery acid, Strong Hydrofluoric Acid
1,000,000	pH = 1	Hydrochloric acid secreted by stomach lining
100,000	pH = 2	Lemon Juice, Gastric Acid Vineger
10,000	pH = 3	Grapefruit, Orange Juice, Soda
1,000	pH = 4	Acid rain / Tomato Juice
100	pH = 5	Soft drinking water Black Coffee
10	pH = 6	Urine / Saliva
1	pH = 7	"Pure" water
1/10	pH = 8	Sea water
1/100	pH = 9	Baking soda
1/1,000	pH = 10	Great Salt Lake Milk of Magnesia
1/10,000	pH = 11	Ammonia solution
1/100,000	pH = 12	Soapy water
1/1,000,000	pH = 13	Bleaches Oven cleaner
1/10,000,000	pH = 14	Liquid drain cleaner

pH Scale

The Molecules of Life

There are four major groups of molecules that make up living organisms: **carbohydrates**, **lipids**, **proteins**, and **nucleic acids**. These macromolecules make up cellular structures and direct and contribute to all life functions: reproduction, metabolism, growth, development, organization, and homeostasis. Each of the four types of organic biomolecules is composed of small molecules that can link together with other small, related molecules to produce large molecules. The smallest individual molecule in each group is generally called a monomer, when two molecules bond together they become a dimer, and when more than two molecules of one group are bonded together, they are called a polymer.

The process of joining together monomers of carbohydrates, proteins, and nucleic acids is called polymerization. In these groups of biomolecules, a hydrogen (H+) from one monomer breaks away to bond with a hydroxide ion (OH–) that has split from another monomer. A water molecule is released and the two monomers are joined to form a dimer. This process is called **dehydration synthesis** because water is removed. The opposite reaction, in which water is added to a dimer or polymer and the larger molecule is split apart, is called a **hydrolysis reaction**.

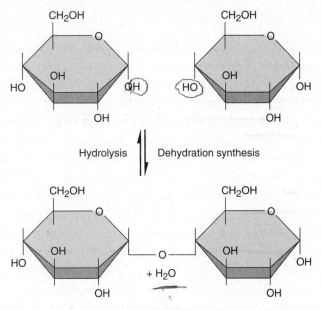

Hydrolysis and Dehydration Synthesis

Carbohydrates

Carbohydrates, or sugars and starches, are one of the four organic (carbon-based) macromolecules that keep living things alive. The cell's main food source, providing the chemicals that cells use for energy or fuel, carbohydrates can provide long-term food storage and can be used by cells as structural components. Carbohydrate molecules are made of carbon (C), hydrogen (H), and oxygen (O) atoms. Carbohydrates are composed of a carbon backbone, with two hydrogen atoms for every oxygen atom.

There are many different types of carbohydrates, but glucose ($C_6H_{12}O_6$) is the most common. The smallest carbohydrate monomers are called monosaccharides, this group includes glucose, galactose, and fructose. Common dimers, or disaccharides, include sucrose (made from glucose + fructose), lactose (glucose + galactose), and maltose (glucose + glucose). Polysaccharides contain many covalently bonded smaller sugar molecules and are sometimes called complex carbohydrates. Common polysaccharides include cellulose (found in plants, this structural molecule gives plant cells rigidity and strength, allowing them to grow tall), glycogen (provides fuel storage in animals), and starch (provides fuel storage for plants).

Naturally made by all photosynthetic organisms, plants provide many of the carbohydrates that we eat. Carbohydrates are found in all fruit and vegetables; in the grains that make up cereals, pasta, and bread; and in rice and potatoes. In fact, the part of the potato that we eat is part of the plant's root system that stores starch for the plant to break down later and use as food for itself. There are photosynthetic organisms that are not plants, such as algae, that provide us with thickening agents for pudding and ice cream and with wraps for our sushi, and the blue-green bacteria *Spirulina*, which can be found in some health foods.

Lipids

Lipids, including oils (liquid), fats (solids), and waxes, are used as nutrient storage molecules and as a main component of cell membranes. Lipid molecules are made of carbon (C), hydrogen (H), and oxygen (O) atoms. This may sound to you like a carbohydrate, but the atoms are arranged differently in lipids. Lipids tend to be large molecules and have many more hydrogen atoms than oxygen atoms.

There are several types of lipids. Fats are composed of fatty acid monomers, which are made up of long, linear chains of hydrocarbons (CH_2) attached to a terminal (at the beginning of the chain) carboxyl group (COOH). If, along the length of the chain, each carbon has two hydrogens bonded with it, the chain is considered to be saturated. Saturated fats are found in animal products and are solid at room temperature. Saturated fats and trans-fats, which are processed to be more useful in a commercial application, are less healthy to eat because they can raise the blood concentration of harmful fats. Mono- and poly-unsaturated fats can lower blood concentration of unhealthy fats. The main unhealthy fats in the blood are triglycerides and LDL (low density lipoprotein) cholesterol. A healthy blood fat is HDL (high density lipoprotein) cholesterol; it reduces LDL cholesterol levels. If the fatty acid chain has any double bonds between the carbons in the chain, and therefore, the carbons wouldn't be bonded to two hydrogens, then it is called an unsaturated fat.

Fatty acids may be used as nutrients by a cell or they may be combined with other molecules to become a structural component of the cell. The two most common places to find fatty acids are bonded to the monomer, glycerol; in triglycerides, as a major energy reserve for the body; and in phospholipids, as the most common component in cell membranes. Another major group of lipids is the steroids. Steroids do not contain fatty acids and, instead of being linear, are arranged in rings. Cholesterol is a common steroid found in cell membranes and is an ingredient in other hormones.

Lipids are found in meats, dairy products, nuts, olives, oils, and mayonnaise. While we want to be careful with the fat content of our diets, lipids are important to the healthy function of the cells in our bodies. If we understand the qualities of the types of lipid we ingest, we will be able to make healthy decisions about what we eat.

Proteins

Protein is a major, and diverse, component of cells, tissues, organs, systems, and organisms. Proteins are structural elements in every cell and are involved in every type of cellular function. Elements that make up proteins are carbon, hydrogen, oxygen, and nitrogen (N). Every protein has a specific type of job to do, based on the way it is put together and on the DNA instructions for assembling it. Proteins tend to be very large and complex molecules, but they are made up of simple monomers called **amino acids**. When two amino acids are joined together, they become a peptide and are attached by a peptide bond. When many amino acids are attached together, they become a polypeptide, or a protein.

Proteins are assembled, as needed, by all living things according to DNA instructions that are carried out along the ribosomes in the cytoplasm of every cell. DNA directs protein assembly by communicating the order, via mRNA, in which amino acids are put together. There are only twenty different amino acids, but they can be arranged in hundreds of thousands of patterns. As proteins are assembled, they are not only joined from one amino acid to the next, but are folded, twisted, and bonded into very particular three-dimensional shapes that allow them to do their jobs perfectly. So, even if the sequence of amino acids is the same, the resulting protein may have a different three-dimensional shape and work differently.

Proteins do many jobs. Structural proteins provide strength and support to cells and body parts. These proteins tend to be tough, flexible, and stringy and include collagen, fibrillin, elastin, and keratin. Ligaments, tendons, the membrane surrounding our arteries, and the tissue of our heart valves are all strengthened by connective tissues that contain collagen, fibrillin, and elastin. Keratin adds toughness and strength to the horns of a goat, the hooves of a horse, the beak of a bird, and to your fingernails, toenails, and hair. Another group of structural proteins helps our muscles move, these are called contractile proteins. Actin and myosin are contractile proteins found in our muscle cells; they pull against each other to contract one muscle while relaxing in an opposing muscle, allowing us to have movement.

Some proteins help other molecules to do their jobs. Transport proteins can help molecules pass in and out of a cell by moving them through the cell membrane. Transport proteins called cytochromes are important during cellular respiration because they move electrons through the electron transport chain. The hemoglobin in your blood is a transport protein with embedded iron that carries oxygen to your cells. Storage proteins hold onto amino acids so they are available when needed by an organism. Have you ever heard that vegetarian diets tend to contain a lot of dried beans? Well, these plants belong to a group called legumes that are loaded with storage proteins and amino acids. Legumes, nuts, and soy products are an excellent non-meat, dietary source of protein. Storage proteins (and their amino acids) are also found in mammalian milk (the protein is called casein) and in egg whites (a protein called ovalbumin).

Some hormones are proteins. Hormones are chemicals that are released from one part of the body and have an affect on another part of the body. Hormonal proteins help to coordinate specific activities of the body. Insulin is a hormonal protein that regulates the concentration of sugar in our blood by controlling glucose metabolism. Uterine contractions during childbirth are regulated by the hormonal protein, oxytocin. Vasopressin is a hormonal protein that prevents dehydration as it regulates the reabsorption of water by the kidneys. Vasopressin causes blood vessels to constrict, or tighten, and in doing so, can cause blood pressure to rise. Somatotropin is a protein made in the pituitary gland which promotes muscular protein production, acting as a growth hormone.

Antibodies are proteins produced to assist the immune system in fighting off invading materials, like bacteria or viruses. They are made by white blood cells called lymphocytes. Also known as immunoglobulins, antibodies circulate in the blood and body fluids of an animal, defending the organism from foreign materials

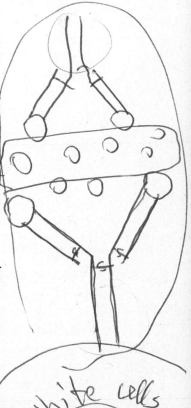

called antigens. Once antigens are immobilized by antibodies, white blood cells can come in and destroy the foreign invaders.

Enzymes are a special class of proteins that help chemical reactions to progress. If you wanted to cause a chemical reaction in a science lab, you would use heat and a mixer to speed up the reaction. We don't turn up the heat or turn on a mixer in our bodies. Instead, enzymes act as **catalysts** to promote the chemical reactions that keep us alive. Many enzyme names end with the suffix –ase. Amylase is an enzyme that speeds up the breakdown of complex sugars and starches in our mouths. A glass of milk contains the sugar lactose, if you drink the milk, an enzyme called lactase should trigger a rapid breakdown of the disaccharide lactose into galactose and glucose. If you are unable to metabolize the lactose because your body lacks this enzyme, you are considered to be lactose intolerant. You will probably develop an upset stomach and may want to avoid dairy products in the future.

Nucleic Acids

Nucleic acids are the final group of the four organic macromolecules. There are two major categories of nucleic acid, deoxyribonucleic acid (DNA) and ribonucleic acid (RNA). Unlike the other groups of macromolecules that provide structure and energy, DNA and RNA are used exclusively by cells for controlling and directing cellular activity. The elements that make up nucleic acids are carbon, hydrogen, oxygen, nitrogen, and phosphorus.

What is DNA and why does everyone make such a fuss about it? DNA is a long, double helix, meaning that it looks like two old-fashioned telephone cords coiled together. It's made up of thousands of monomers called **nucleotides** that are made of a sugar molecule called deoxyribose, with a phosphate group attached on one corner and a nitrogen base molecule attached on another corner. When it's uncoiled, the structure of DNA looks like a ladder. Along the side rails are simply alternating sugar (deoxyribose)-phosphate-sugar-phosphate molecules. The alternating sugar-phosphate molecules are always identical to each other. These side rails are strong yet flexible and give the different nitrogen bases a place to hold on to and so not to lose their correct place in the sequence of thousands of other nitrogen bases.

So, the order of the nitrogen bases is what makes DNA so powerful and important. DNA has just four nitrogen bases: the pyrimidine bases cytosine (C), thymine (T), and the purine bases adenine (A), and guanine (G). These bases work in pairs, one purine plus one pyrimidine to form the rungs, or the steps, of the ladder. Specifically, in DNA, adenine only binds with thymine and cytosine only binds with guanine. Adenine's bond with thymine is the weaker of the two because it is a double hydrogen bond, while cytosine has a stronger triple hydrogen bond with guanine. DNA has four types of nucleotides and they are named after the first letters of their nitrogen base groups: A, C, T, and G.

As we continue reviewing nucleic acids, keep in mind that the power and control of DNA comes from the order of adenine, thymine, cytosine, and guanine along one side of the DNA ladder. Sections of these sequences (anywhere from less than one-hundred to several thousand nucleotides) are called genes, and they direct the assembly of a specific protein as it is needed. A whole, long strand of DNA, including many genes as well as many unusable sequences of nitrogen bases, is called a chromosome.

We still have not discussed the second group of nucleic acid, ribonucleic acid. RNA is also made up of a sequence of nucleotides; it has the same plan as DNA, but there are several important differences. First, RNA uses the sugar ribose, instead of deoxyribose, to build its backbone. It uses the same phosphate group as DNA and builds the equivalent sugar-phosphate-sugar-phosphate sequence. Second, the nitrogen base groups match up almost exactly like DNA except that RNA uses uracil instead of thymine to bond with adenine. The third difference is that RNA is a single-stranded molecule. Fourth, DNA stays primarily in the nucleus, while the RNA travels out of the nucleus and spends most of its time in the cytoplasm.

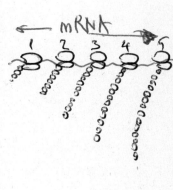

The fifth, and final, major difference is that RNA comes in several forms, as directed by the DNA, in order to perform different functions: messenger RNA, ribosomal RNA, and transfer RNA. Messenger RNA (mRNA) is assembled to complement, or match, a needed gene on a DNA strand. It then takes that matching sequence of nucleotides to a ribosome and uses the nitrogen base sequence to attract amino acids coded for by that sequence. Ultimately, the amino acids are bonded to one another in the order needed to make the correct protein as originally directed by the DNA. Ribosomal RNA (rRNA) becomes part of the ribosome, the organelle where proteins are formed. Transfer RNA (tRNA) travels through the cytoplasm and brings one amino acid to the ribosome where it's needed to match a small part of the amino acid sequence originally dictated by the DNA. Each tRNA molecule carries one type of amino acid and is reusable, as soon as it drops off one amino acid it goes back out, picks up another, and drops off the next one.

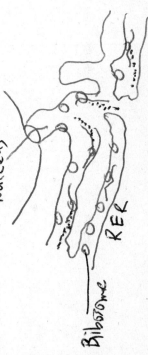

So That's a Taste of Biochemistry

Can you now see why we need to understand how biomolecules work in order to understand how living things stay alive? In order to perform activities, we need to get energy (from carbohydrates and lipids) and parts to make new cells (structural proteins), and, since energy and spare parts are packaged up in our food, we must metabolize (using enzyme proteins) the food to access its nutrient reservoir. We are also designed to store some nutrients (complex carbohydrates and lipids) in case food becomes less available. In order to replenish old and damaged cells, we must have both a collection of parts (structural proteins) and a set of directions (nucleic acids). In order to follow the directions, we need a fluid medium (cytoplasm) to help move some of the instructions (mRNA) to other parts of the cell.

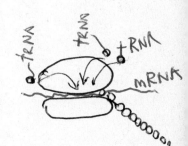

Important Terms to Remember

Amino acid: Building blocks of proteins. There are 20 different amino acids that are directed by DNA to be arranged into millions of three-dimensional arrangements to form all of the proteins needed by all living things.

Atom: The smallest functional unit of an element; made up of protons, neutrons, and electrons.

Biochemistry: The science of how chemistry and biology impact each other; understanding the chemical processes that occur in living things.

Bonding: The attachment of elements to one another through chemical attractions.

Carbohydrate: Includes all sugars and starches; one of the four major organic macromolecules, made of carbon, hydrogen, and oxygen with a ratio of two hydrogen atoms to each oxygen atom. Used as a source of cellular energy and energy storage; monosaccharides, disaccharides, and polysaccharides; produced during photosynthesis.

Catalyst: An enzyme that reduces the activation energy for a chemical reaction; allows chemical reactions to occur in living organisms without altering normal body functions or temperature.

Covalent: Bond created by the sharing of electrons in the outermost electron orbital.

Dehydration synthesis: Chemical reaction in which two molecules are joined together as a molecule of water is removed.

Element: Substance made up of only one type of atom.

Hydrogen: Bond created by the attraction of a positively charged hydrogen atom to a negatively charged atom (often oxygen or nitrogen).

Hydrolysis reaction: Chemical reaction in which one molecule is split apart by the addition of a molecule of water.

Ionic: Bond created by the giving and taking of electrons in the outermost electron orbital.

Lipid: Oils, fats, and waxes; one of the four major organic macromolecules. Used for energy storage and as an integral cell membrane component; made of carbon, hydrogen, and oxygen, with more than two hydrogen atoms per every oxygen atom; includes fatty acids, glycerol, steroids.

Mixture: A blend of chemicals in which the molecules being combined do not join together or change in structure.

Molecule: A group of atoms with two or more different elements held together by bonds.

Nucleic acid: DNA and RNA; one of the four major organic macromolecules; made of carbon, hydrogen, oxygen, nitrogen, and phosphorus; its monomer is a nucleotide.

Nucleotide: Building block of nucleic acids; made up of a sugar, a phosphate group, and a nitrogen base group (adenine, cytosine, guanine, thymine, or uracil).

Orbital: The part of an atom outside of the nucleus where electrons circulate. There are different levels which, when full with their maximum number of electrons, provide the atom with stability.

Peptide: Bond created between amino acids to form proteins.

pH: A scale of 0 to 14 by which acidity and alkalinity are measured; a strong acid measures 0, a strong base measures 14, and pure, neutral water measures 7 on the pH scale.

Protein: One of the four major organic macromolecules, made of carbon, hydrogen, oxygen, and nitrogen; used for cellular structure and contraction, as enzymes, antibodies, and hormones; made of many amino acids bonded together into three-dimensional shapes.

Solution: Molecules are mixed together and they join together chemically; cannot be easily separated.

Practice Questions

Multiple-Choice

1. You are looking at a Periodic Table hanging on the wall in science class. In each box you see either one capital letter or a capital letter followed by one or more lowercase letters. These letters are each symbols representing a type of _____.

   (A) atom
   (B) element
   (C) molecule
   (D) formula

2. How many sugar molecules are in the formula $6C_6H_{12}O_6$?

   (A) 1
   (B) 6
   (C) 24
   (D) 144

3. In biology class, you've been discussing cellular metabolism and the process of breaking large molecules down to small enough molecules to pass through cell membranes. This process is called _____.

   (A) hydrolysis
   (B) dehydration synthesis
   (C) dimerization
   (D) carbon fixation

4. The tree growing outside your window is towering way above the house. Your uncle said he thinks it is about 100-feet tall. You're wondering, if it doesn't have a skeleton (and you know from biology class that it doesn't), how can it grow so tall and straight? It's got to be heavy, why doesn't it just fall over or collapse in on itself?

   (A) Its cells contain a polysaccharide called cellulose that makes it strong.
   (B) Its cells contain a polysaccharide called glycogen that makes it strong.
   (C) Its cells contain a structural protein called keratin that makes it strong.
   (D) Its cells contain a steroidal lipid called collagen that makes it strong.

5. A genetic disorder called Muscular Dystrophy causes human muscles to stop working properly. Muscular Dystrophy happens because a gene, on one strand of a parent's DNA, has an incorrect sequence of nitrogen bases. This sequence directs a necessary protein to be made incorrectly and that protein will not do its job well. Which proteins are likely to be affected in Muscular Dystrophy?

   (A) oxytocin and vasopressin
   (B) collagen and keratin
   (C) amylase and lactase
   (D) actin and myosin

6. The genetic disorder phenylketonuria, PKU, occurs when an enzyme doesn't do its job of breaking down an amino acid called phenylalanine. Newborn babies in the United States are tested for this genetic defect, and, if it's found, children must remain on a strict diet for their whole life. If a child with PKU does not stay on this diet, phenylalanine accumulates in and becomes toxic to the child's brain and nerves, ultimately causing brain damage. What is the normal function of the protein that causes PKU?

   (A) It is a catalyst.
   (B) It is an immunoglobulin.
   (C) It is a hormone.
   (D) It is a transport protein.

7. You are working in a biochemistry lab examining a slime mold that was growing on a decomposing tree in a rain forest. You are able to establish that the molecules you've extracted from this organism contain the elements carbon, oxygen, hydrogen, and nitrogen. What kind of biomolecule do you think you're looking at?

    (A) carbohydrate
    (B) protein
    (C) lipid
    (D) nucleic acid

8. You are still working with that slime mold and you've taken some new cellular material that contains carbon, oxygen, hydrogen, nitrogen, and phosphorus, and your latest chemical analysis indictates molecules of uracil. What kind of biomolecule do you think you're looking at?

    (A) lipid
    (B) protein
    (C) RNA
    (D) DNA

9. Your boss in the lab asks you to list the differences between DNA and RNA. Of the following, which does she tell you is incorrect?

    (A) DNA is single stranded and RNA is double stranded.
    (B) DNA remains primarily in the nucleus while RNA spends a lot of time in the cytoplasm.
    (C) DNA contains thymine while RNA contains uracil.
    (D) DNA has the sugar deoxyribose and RNA contains ribose.

10. It's been a long week in the lab, and you and your coworkers decided to go out for pizza. Which of the following is not a major nutrient source in your pizza?

    (A) carbohydrate
    (B) protein
    (C) lipid
    (D) nucleic acid

Free-Response

1. Your best friend has decided to become a vegetarian and has asked you to help develop ideas for a healthy vegetarian diet. What might you suggest to your friend?

2. Your friend's newborn brother has been diagnosed with a genetic disorder called phenylketonuria, PKU. You and your friend are trying to think of other ways to remedy this genetic disorder that would not require him to stay on a restricted diet for his whole life. You've just logged on to the school's science database. What alternatives might you begin to research?

3. You are working in a lab and have been given a tiny specimen to evaluate. You have been told only that it is a biological specimen containing a single type of macromolecule. Before you begin any extensive testing, what information do you need in order to ultimately determine whether this is a protein, carbohydrate, lipid, or nucleic acid?

Answer Explanations

Multiple-Choice

1. **B** The symbols on a Periodic Table represent elements. Atoms make up elements, molecules are made up of two or more elements, and a formula explains how molecules interact.

2. **B** The number "6" in front of $6C_6H_{12}O_6$ tells us that there are six molecules. If there was one molecule the formula would simply be read $C_6H_{12}O_6$. In one molecule of this sugar, there are 24 atoms $(6 + 12 + 6)$ and in six molecules of this sugar are 144 atoms (6×24).

3. **A** Hydrolysis is the reaction in which a molecule of water is added in order to break apart a larger molecule. Dehydration synthesis results in the joining of two monomers. Dimerization and carbon fixation have nothing to do with breaking down large molecules.

4. **A** The polysaccharide called cellulose gives strength to the cell walls of plants. Glycogen, keratin, and collagen are all found in animals and not in plants.

5. **D** Actin and myosin are contractile proteins that, when they are healthy, make our muscles work correctly. Oxytocin and vasopressin are hormonal proteins. Collagen and keratin are structural proteins that give strength and flexibility to our ligaments, tendons and fingernails. Amylase and lactase are enzyme proteins that help us to metabolize sugars and milk products.

6. **A** It is a catalyst; it causes a chemical reaction to occur and even to speed up under normal body conditions. It doesn't fight off antigens, send messages, or take materials to distant parts of the cells or the body.

7. **B** It appears to be a protein. It would contain carbon, oxygen, and hydrogen if it were a carbohydrate or lipid, and if it were a nucleic acid, it would contain phosphorus in addition to the carbon, oxygen, hydrogen, and nitrogen that you've already found.

8. **C** It is the nucleic acid, RNA. You know this because it contains the elements present in nucleic acid and because RNA has uracil while DNA does not.

9. **A** The incorrect answer is that DNA is single-stranded and RNA is double-stranded. DNA does primarily stay in the nucleus and contain thymine and deoxyribose. RNA does work in the cytoplasm and it contains uracil and ribose.

10. **D** Nucleic acids are not a major food group although they do provide our diet with breakdown products such as phosphorus. Carbohydrates, proteins, and lipids are major nutrient sources.

Free-Response Suggestions

1. **Your best friend has decided to become a vegetarian and has asked you to help develop ideas for a healthy vegetarian diet. What might you suggest to your friend?**

   Any healthy diet needs to include adequate nutrients from all of the food groups. The group that may be missing from your friend's diet as a vegetarian would be proteins. Proteins are the main structural component of the animal muscles that we call meat. When we eat meat, we metabolize, or break down, the proteins in the meat to its building blocks, amino acids. We then rearrange those amino acids into the proteins that our bodies need. Your friend needs to find those amino acids in non-meat forms; these include legumes and bean products, nuts, and soy products.

2. **Your friend's newborn brother has been diagnosed with a genetic disorder called phenylketonuria, PKU. You and your friend are trying to think of other ways to remedy this genetic disorder that would not require him to stay on a restricted diet for his whole life. You've just logged on to the school's science database. What alternatives might you begin to research?**

   You and your friend just logged on to the school's science database to research alternatives to a restricted diet for PKU patients. Perhaps you'll look for information about whether anyone has investigated an artificial or supplemental enzyme product to replace the enzyme that isn't breaking down phenylalanine the way it should. Maybe you'll look to see if there has been any successful gene therapy to encourage the cells of PKU patients to make this enzyme correctly.

3. **You are working in a lab and have been given a tiny specimen to evaluate. You have been told only that it is a biological specimen containing a single type of macromolecule. Before you begin any extensive testing, what information do you need in order to ultimately determine whether this is a protein, carbohydrate, lipid, or nucleic acid?**

   So, is the sample a protein, carbohydrate, lipid, or nucleic acid? Well, you can assume the sample will contain carbon, hydrogen, and oxygen regardless of which biological molecule it is. Why not test first for the presence of phosphate?

If phosphate is present in relatively large amounts, the sample is likely to be a nucleic acid. If there is little or no phosphate but plenty of nitrogen, you are probably looking at a protein. If there is no nitrogen either, you may want to consider the hydrogen-to-oxygen ratio. If there is clearly more than twice as much hydrogen as oxygen, the sample is likely to be a lipid. If the ratio is closer to two hydrogens for every oxygen, you're probably looking at a carbohydrate.

CHAPTER 4

Transport in the Cell

Outline for Organizing Your Notes

I. Structure and Function of the Cell Membrane
   A. Phospholipid Bilayer
   B. Transport Proteins
   C. Cholesterol

II. Passive Transport
   A. Doesn't Require the Cell to Use Energy
   B. Diffusion
      1. Dynamic Equilibrium
      2. Osmosis
      3. Solutions: Hypotonic, Isotonic Hypertonic
   C. Facilitated Diffusion
      1. Channel Proteins
      2. Enzymes

III. Active Transport
   A. Requires the Cell to Use Energy to Move Molecules in Difficult Situations
      1. Unit of Energy: Adenosine Triphosphate (ATP)
      2. Large Molecules
      3. Strong Charge
      4. Movement Against the Concentration
   B. Types of Active Transport into the Cell
      1. Endocytosis: Pinocytosis, Phagocytosis
   C. Active Transport Out of Cell
      1. Exocytosis

Why is it important to understand how the cell selects what will cross in and out of its boundaries? We humans make choices constantly, why is it such a big deal that cells can do the same thing? Well, we humans have a brain that reviews our options and makes our choices for us. Cells don't have a brain to make choices regarding what will enter and leave. You may have heard someone say that a cell's nucleus is its "brain," but that's not an accurate description. The nucleus controls the reproduction and many other activities of the cell through chemical messages, but it doesn't make "choices," and it doesn't control transport across the cell membrane. The cell membrane is the cellular structure that selects what will cross in and out of the cell.

The Structure and Function of the Cell Membrane

The cell membrane, also known as a plasma membrane, is the structure that encloses the cytoplasm and most cellular organelles. It's described as a **phospholipid bilayer** and as a **selectively permeable membrane**. A common representation of the plasma membrane is called a **fluid mosaic model**. These descriptions reflect a number of characteristics of cell membranes; the membrane is constantly and smoothly in motion (it's "fluid"), it looks like many little round tiles pieced together into a pattern (mosaic), its main ingredients are phosphate groups and lipid molecules (phospholipid), they exist together in two layers (bilayer), and they are able to choose which molecules will be allowed to pass through (selectively permeable). Other important structures in a cell membrane include an assortment of proteins: transport proteins to help large molecules, or molecules that have difficulty crossing in or out, signaling proteins to assist communication between cells, and marker proteins to help cells recognize each other. Cholesterol molecules have an important role in healthy membrane function because they maintain stability by holding the two phospholipid layers together.

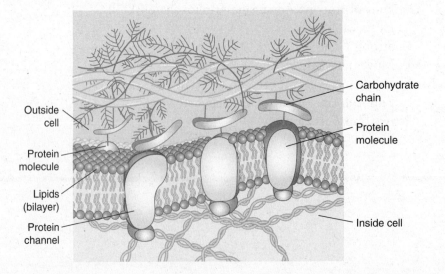

The cell membrane is also important as an anchoring site. It maintains cellular shape and support by providing the cytoskeleton with a place to attach to and it gives extracellular organelles like cilia and flagella a base for linkage and support for mobility. In multicellular organisms, the cell membrane provides a surface for attachment of the extracellular matrix, allowing cells to work together to form tissues. Sometimes, in plants, fungi, some protists, and some bacteria, the cell membrane is surrounded by a cell wall that is essentially used for strength and support, not for selective permeability. The cell membrane is quite unique in its selective permeability.

An Analogy

The cell membrane is very similar to security at an airport. It exists along the outer boundaries of the airport terminal and carefully screens everyone who wants to get inside to the functional parts of the airport. As we proceed through this unit on cellular transport, try to think of how each cell membrane structure relates to this airport security analogy.

How Are Structure and Function Related?

So, the first major thing to know about how cell membranes select which molecules can cross is that two layers of phospholipid molecules make up the majority of the membrane. These phospholipid molecules have different chemical properties at their two ends: the hydrophilic (water-loving) head is polar and the hydrophobic (water-hating) fatty acid tails are non-polar. This means that the (polar) head will be attracted to water and to other mildly electrically charged molecules. Since the cytoplasm (intracellular fluid) and extracellular fluid are water-based, they are attracted to the polar heads. The polar heads, therefore, make up the surfaces of the cell membrane that are in direct contact with cytoplasm and extracellular fluid. The hydrophobic fatty acid tails extend inward into the membrane and away from the polar heads.

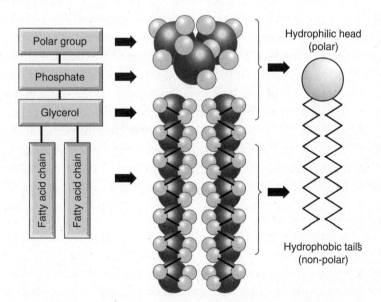

> **An Analogy**
>
> Travelers are similar to the molecules that want to cross the cell membrane, integral proteins can be considered gates through which molecules travel. The phospholipid molecules are like the multitude of security workers who check our ID and boarding passes. They include the transit police officers who prevent people from parking their cars for too long outside the building. They also include the skycaps and redcoats, airline employees who are constantly assessing the identity and the safety of all the passengers lining up or milling about trying to get to and through the security checkpoint gates.

What Is Cellular Transport?

All different kinds of materials need to get into or out of a cell's cytoplasm. Cells need sodium (Na), chlorine (Cl), potassium (K), magnesium (Mg), Iron (Fe), calcium (Ca), and many other elements. Some of these elements need to cross the cell membrane as charged particles because they have given up or taken electrons. These elements are then called ions. Ions can be attracted to, or repelled by, the polar heads of the cell membrane. A small, mildly charged, ion is likely to pass easily through the cell membrane. Large particles, molecules that don't interact easily with water, and those with strong charges, may need some help to pass through the membrane.

> **An Analogy**
>
> Different travelers are going to have different needs as they move through the airport and even through security. Some people aren't planning to travel, they are just accompanying someone who will be traveling; these people will be turned away at the security gate. Some travelers arrive at the security check point with minimal baggage, their documents in order, their shoes off, and their laptops out of the bag. Other travelers have to be coached through this process. Some travelers have to be screened more carefully than others. Some travelers require physical assistance moving, and so an airport customer service agent may need to help them through security or may use a cart to drive them through the airport. Passengers may require verbal instructions to proceed through the airport or they can even be denied entry to the airport. All these scenarios are similar to the variety of situations encountered by molecules as they are trying to cross through a cell membrane.

Passive Transport

Some molecules just slide through the cell membrane. Other molecules pass through the cell membrane easily because they are moving along a concentration gradient, from an area where they are in high concentration to an area where they're in a lower concentration. This process doesn't require the cell to use energy, and as a result, the process is called **passive transport**. During passive transport, molecules may simply squeeze between the phospholipid molecules. Other mechanisms for passive transport require the assistance of some types of transport proteins.

The process of moving with a concentration gradient, from high to low concentration, is called **diffusion**. An example of diffusion in everyday life would be the way that fragrance moves through a closed room when someone sprays an air freshener. At first, the smell is strong near where it was sprayed and it cannot be smelled on the other side of the room. But soon the fragrance has dispersed throughout the room; it is less strong where it was first sprayed but now it can be smelled across the room. As the fragrance molecules spread around, they are reaching a condition called **dynamic equilibrium**, the point where the number of fragrance molecules is equal throughout the room. In a cell, diffusion across the cell membrane can help to move molecules toward dynamic equilibrium between the inside and outside.

Because a cell membrane allows specific molecules to cross and prevents other molecules from crossing, it is called a **selectively permeable membrane**. This membrane allows water to cross into or out of a living cell unless it's prevented by an opposing force. Water, a solvent, will move across the membrane to keep the concentration of other molecules, solutes, at dynamic equilibrium. This process, **osmosis**, works because the water dilutes, or concentrates, the ions or molecules it contains to the appropriate concentration for that cell in that environment. When a healthy cell is placed into a **hypotonic** solution the concentration of solutes is higher inside the cell than in the surrounding solution. Water will enter the cell; it will swell and may ultimately burst. When a cell is placed into **hypertonic** solution, water will move out of the cell because there is a higher concentration of solutes outside than inside the cell. Under these circumstances, the cell will shrivel up and may die. A healthy cell placed into **isotonic** solution will maintain its normal size and shape because the concentration of solutes in and outside the cell is in balance, and it has reached dynamic equilibrium.

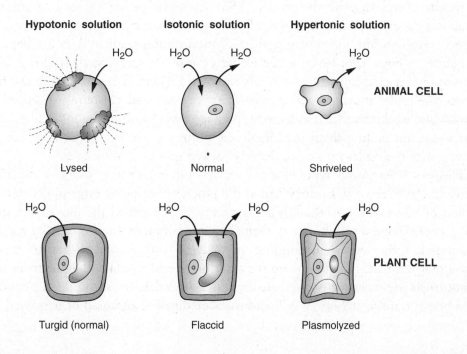

Hypotonic solution	Isotonic solution	Hypertonic solution	
H_2O	H_2O H_2O	H_2O	**ANIMAL CELL**
Lysed	Normal	Shriveled	
H_2O	H_2O H_2O	H_2O	**PLANT CELL**
Turgid (normal)	Flaccid	Plasmolyzed	

Sometimes a cell needs a molecule or an ion that cannot pass easily through the cell membrane. In this case, the cell membrane uses its **transport proteins** to help move the molecules in a passive transport process called **facilitated diffusion**. Molecules are still moving with the concentration gradient, so the use of energy isn't required in facilitated diffusion. The transport proteins used in this type of passive transport are channel proteins, which provide a tunnel or tube for the molecules to travel through, and enzymes that link with specific passing molecules and pull them through the membrane.

An Analogy

Passive transport is similar to airport security in that many passengers can pass through simply and without really using much energy (but do remember that any energy use at all by the cell does make it an active process). Like passively transported molecules through protein channels, many travelers just slide through the security gate, without any hassle.

Active Transport

Molecules that still can't cross the cell membrane need more help and this help comes at a cost to the cell. A cell must use energy if it has to assist material across the membrane. When a cell uses energy to move materials across the plasma membrane, the process is considered to be **active transport** because the cell is actively working to move its load. Energy available to the cell is contained in the high energy bonds of a molecule called adenosine triphosphate (ATP).

There are several reasons why a cell would need to actively transport materials across its membrane. Movement against the concentration gradient, as well as transport of strongly charged and large molecules, are all situations that are likely to require energy to drive the process. Membrane pumps use energy to transport some large molecules and some ions from areas of low concentration to areas of higher concentration. Movement against a concentration gradient does not happen in nature without help. Imagine the earlier example of spraying air freshener and the way that it randomly diffuses across the closed room. Can you picture the fragrance molecules moving back toward the spray can and clustering together? It would take work to make that happen, perhaps a fan blowing the molecules back and a vacuum pulling them in. Membrane pumps act in this way; they actively move sparse molecules into more densely packed groups of molecules.

There are several other types of active transport that use energy to move materials into and out of the cell. **Endocytosis** is the process of bringing large molecules, or groups of molecules, into the cell using energy (ATP) to fuel the movement, and using a containment vessel to carry them. The containment vessel is called a vesicle. The vesicle is formed by surrounding the material to be moved with a cell membrane that ultimately carries it into the cell during endocytosis. **Phagocytosis** and **pinocytosis** are forms of endocytosis in which materials are surrounded by a cell membrane, transported into a cell, and then are digested and used or destroyed by the cell. Phagocytosis brings large, solid particles into the cell while pinocytosis brings in droplets of liquid.

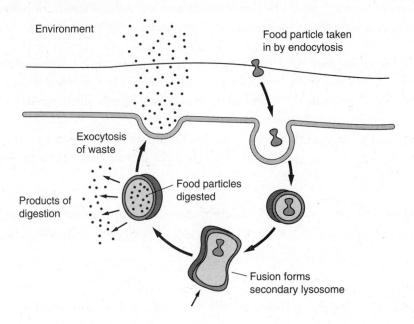

Active transport out of cell is called **exocytosis**. Exocytosis is the reverse of endocytosis; it involves a vesicle that is filled with waste or unneeded materials and its movement from inside the cell toward the cell membrane. The vesicle membrane fuses with the cell membrane then opens up on the cell's exterior to discard the waste.

An Analogy

What do you think active transport relates to in our ongoing comparison? How about our plane trip? We are encapsulated in a plane, we travel to a destination, we have our vacation (or whatever we traveled for), we travel back, we disembark, leave the airport through the reverse side of the security gates, and then go home again. Granted, we don't get digested in the process but it is a lot like endocytosis and exocytosis in terms of being transported to and from a destination, across a physical barrier, while confined inside a vessel.

Important Terms to Remember

Active transport: Movement across the cell membrane usually using adenosine triphosphate (ATP) as the unit of energy; materials move against the concentration gradient (from an area of low concentration into an area that already has a higher concentration).

Bilayer: Two layers.

Diffusion: Movement of molecules from an area of high concentration to an area of lower concentration (movement is with the concentration gradient).

Dynamic equilibrium: As molecules scatter, and randomly bounce off of each other, they reach a point where all the molecules are evenly dispersed in an enclosed area. Molecules continue to move but are ultimately evenly distributed.

Endocytosis: Active transport into a cell; materials are surrounded by a membrane to form a vesicle and are transported into the cells intracellular fluid; phagocytosis and pinocytosis are examples.

Exocytosis: Active transport out of cell; materials are surrounded by a membrane to form a vesicle and are transported out of the cell and into the extracellular fluid.

Facilitated diffusion: Assisted movement across a cell membrane, moving with the concentration gradient; assistance usually provided by transport proteins.

Fluid mosaic model: Scientifically accepted representation of the cell membrane; this model acknowledges the constantly shifting patterns of phospholipids, proteins, and cholesterols in the surface of the cell membrane.

Hydrophilic: Water-loving: molecules that are attracted by water (such as weak ions).

Hydrophobic: Water-hating; molecules that are repelled by water (such as oils and fats).

Hypertonic: Solution in which there is a higher concentration of solutes than inside the cell. Water will move out of the cell; this may cause the cell to shrivel up and potentially die.

Hypotonic: Solution in which the concentration of solutes is lower than inside the cell; water will enter the cell causing it to swell, the cell could eventually rupture.

Isotonic: Solution in which the concentration of solutes in and outside the cell is equal.

Osmosis: Movement of water across a selectively permeable membrane.

Passive transport: Process doesn't require the cell to use energy; molecules move with the concentration gradient (from an area with a high concentration of solutes into an area with a lower concentration).

Phagocytosis: A type of endocytosis in which solid particles are enveloped in the cell membrane, transported across the membrane into the cell; generally these particles are lysed, or digested, by the cell's lysosomes and available nutrients are then metabolized by the cell.

Phospholipid: A molecule made up of a polar phosphate group attached to non-polar fatty acid tails; a major component of all plasma membranes.

Pinocytosis: A type of endocytosis in which liquid droplets are enveloped in the cell membrane, transported across the membrane into the cell; generally these droplets are lysed, or digested, by the cell's lysosomes and available nutrients are then metabolized by the cell.

Selectively permeable membrane: Cell covering which is able to choose which molecules will be allowed to cross through to the other side.

Transport proteins: Proteins that help in the movement of molecules through the cell membrane; includes channel proteins.

Practice Questions

Multiple-Choice

1. Your friend, Becca, has a sore throat and swollen tonsils. Her parents can't get her in to see the doctor until tonight. Your mom always tells you to gargle with warm salt water when you start getting a sore throat, so Becca tries this and it seems to help. It helps because the solution is _____, and so it _____ the cells.

    (A) hypertonic; shrinks
    (B) hypotonic; shrinks
    (C) isotonic; numbs
    (D) hypotonic; numbs

2. You are building a model of a cell for your science class. What are you most likely to use to represent the phospholipid bilayer?

    (A) a bowl of gelatin and some jelly beans
    (B) a piece of paper, a ruler, and a pencil
    (C) ping pong balls, straws, and pipe cleaners
    (D) a Ziploc sandwich bag filled with gelatin

3. Which biological macromolecules help to make up the structure of the plasma membrane?

   (A) only lipids
   (B) lipids and nucleic acids
   (C) nucleic acids and proteins
   (D) proteins and lipids

4. You decided to have a cup of hot chocolate after dinner. You heated the milk and added the chocolate, then stirred. What was the result of heating and stirring?

   (A) endocytosis
   (B) exocytosis
   (C) dynamic equilibrium
   (D) osmosis

5. Which of the following statements is not true?

   (A) Passive transport is conducted without using any energy.
   (B) Passive transport happens when molecules are found in differing concentrations.
   (C) Passive transport is sometimes helped by proteins.
   (D) Passive transport just uses enough energy to get the job done, nothing more.

6. Nerve cells release chemicals called neurotransmitters in a process that involves wrapping the neurotransmitters in a cell membrane while they are inside the nerve cell. The neurotransmitter is inside a structure called a(n) _____.

   (A) vesicle
   (B) transport protein
   (C) lysosome
   (D) enzyme

7. The neurotransmitter is moved from inside the cell to the cell membrane where it is released from the cell to enter the extracellular fluid. This process is called _____.

   (A) passive transport
   (B) endocytosis
   (C) exocytosis
   (D) pinocytosis

8. What is happening in this diagram?

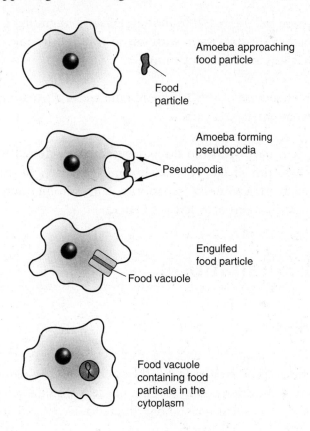

Amoeba approaching
food particle

Food
particle

Amoeba forming
pseudopodia

Pseudopodia

Engulfed
food particle

Food vacuole

Food vacuole
containing food
particale in the
cytoplasm

    (A) exocytosis
    (B) phagocytosis
    (C) pinocytosis
    (D) passive transport

9. Which of the following is not related to passive transport?

    (A) channel proteins
    (B) osmosis
    (C) endocytosis
    (D) diffusion

10. Which of the following is not a function of active transport?

    (A) pinocytosis
    (B) exocytosis
    (C) endocytosis
    (D) facilitated diffusion

Free-Response

1. You were outside last night and found several slugs migrating across your yard. Your mom said to sprinkle them with salt so you did. It looked like they were melting. What was actually happening?

2. How does the structure of the cell membrane create a situation where the cell can choose what enters and exits?

3. Your friend, Jeff, developed the flu and became dehydrated. He went to the hospital and was put on an IV to help him rehydrate. Discuss the types of solutions available (hypotonic, hypertonic, and isotonic) and explain which solution is most likely to help Jeff get better.

Answer Explanations

Multiple-Choice

1. **A** The solution is hypertonic and as a result water will leave the cells of Becca's throat and tonsils. As water leaves the swelling will be reduced and tonsils will shrink. If the rinse was hypotonic the cells would swell up. If the rinse was isotonic it wouldn't make a big difference to the cell size. Numbing is not a function of water or salt concentration.

2. **C** Ping pong balls, straws, and pipe cleaners would be your best modeling tools to represent phospholipid molecules and embedded transport proteins. A bowl of gelatin with jelly beans may better represent cytoplasm, not a plasma membrane. Drawing materials don't make a model, they would make a diagram. A Ziploc bag full of gelatin would not represent a dynamic cell membrane because it's simply a container with material inside.

3. **D** Proteins and lipids are parts of the cell membrane, in phospholipid molecules and transport proteins. Carbohydrates and nucleic acids are not cell membrane constituents.

4. **C** Dynamic equilibrium, the equal distribution of chocolate molecules throughout the milk, was the result of heating and stirring. Making hot chocolate does not involve passive or active transport across the cell membrane.

5. **D** It is not true that "passive transport just uses enough energy to get the job done, nothing more." Passive transport does not use any energy from the cell at all. It is true that passive transport happens when molecules are found in differing concentrations. It is also true that passive transport is sometimes helped by proteins.

6. **A** The structures that carry neurotransmitters to be released from the cell are called vesicles. Transport proteins are structures that help with passive transport. Lysosomes are organelles that digest material for the cell. Enzymes are catalysts, not containers.

7. **C** Exocytosis, a form of active transport, is the process of releasing a material from inside the cell to outside the cell using a vesicle. Endocytosis, including pinocytosis, takes materials into the cell using a vesicle.

8. **B** Phagocytosis is represented in this diagram, the cells ingestion of a food particle using active transport. Pinocytosis, another form of endocytosis, involves the ingestion of a liquid droplet. Exocytosis would be eliminating the food particle.

9. **C** Endocytosis is a function of active transport, not of passive transport. Osmosis, channel proteins, and enzymes are all related to passive transport.

10. **D** Facilitated diffusion is a function of passive transport, not of active transport. Endocytosis, pinocytosis, and exocytosis are all related to active transport.

Free-Response Suggestions

1. **You were outside last night and found several slugs migrating across your yard. Your mom said to sprinkle them with salt so you did. It looked like they were melting. What was actually happening?**

   The cells in the slug are surrounded by a cell membrane that is selectively permeable. When you sprinkle salt on the skin and mucus that covers the slug, there is a higher concentration of salt in the water on the outside of the slug. As a result, water leaves the cells and the body of the slug to try to equalize the salt concentrations on the inside and the outside. The slug loses too much water through dehydration, its body shrivels up, and it finally dies.

2. **How does the structure of the cell membrane create a situation where the cell can choose what enters and exits?**

   The constantly shifting of outermost and innermost layers of phosphate groups attract and repel specific molecules. Those that are attracted will pass through with relative ease. The nonpolar fatty acid tails slide some specific large or strongly charged molecules through. The embedded proteins will interact with specific molecules to pass across the membrane.

3. **Your friend, Jeff, developed the flu and became dehydrated. He went to the hospital and was put on an IV to help him rehydrate. Discuss the types of solutions available (hypotonic, hypertonic, and isotonic) and explain which solution is most likely to help Jeff get better.**

Hypertonic solution is not going to be helpful because it is going to draw even more water out of Jeff's cells and he will get more dehydrated. Hypotonic solution, if it's very mild, might help Jeff because it will push some water into his cells. But this must be done with caution, it can backfire and cells could get too full and burst, a very dangerous situation. Isotonic solution would be most helpful because it's balanced for a normally hydrated person, so with Jeff's dehydration, his cells would get extra water from the isotonic solution until they reached equilibrium.

CHAPTER 5

Cellular Reproduction

Outline for Organizing Your Notes

I. Asexual and Sexual Reproduction

II. The Cell Cycle
   A. Interphase
      1. Gap 1
      2. Gap 0
      3. S Phase
      4. Gap 2
   B. Mitosis
      1. Prophase
      2. Metaphase
      3. Anaphase
      4. Telophase
      5. Cytokinesis
   C. Meiosis
      1. Prophase I
      2. Metaphase I
      3. Anaphase I
      4. Telophase I
      5. Cytokinesis I
      6. Prophase II
      7. Metaphase II
      8. Anaphase II
      9. Telophase II
      10. Cytokinesis II
      11. Gametogenesis
   D. Telomeres
      1. Apoptosis
      2. Immortality
      3. Senescence

How does your skin heal when you get a cut or sunburn? What happens to make skin look different as it ages? Why is cancer so common and why do we hear about it occurring more frequently as people get older? What controls the aging process? Cellular reproduction is the link between all of these questions and their answers. All living things share the characteristic of reproduction. When we think of reproduction we often think of an egg cell being fertilized by a sperm cell to produce unique offspring. It's true; since, babies of different species are being born all the time. But happening with even more frequency is the form of reproduction occurring constantly in every organism, of individual body cells producing two new daughter cells that are essentially identical to the original parent cell, a form of cellular reproduction.

Asexual and Sexual Reproduction

There are different reasons why cells reproduce. When you get a paper cut on your finger, skin cells undergo a form of **asexual reproduction**, called regeneration, to fill in the space and heal your injury. As your puppy grows up, many of its body cells grow and divide so that it can get bigger. In the springtime, when a daffodil plant emerges from its underground bulb, its cells are growing and dividing very quickly in order to grow tall and produce a flower before its growing season is over. Cells in every type of organism are genetically programmed to grow and divide so that the organism can also grow and stay alive. Cell division, the process of making duplicate cells from one original parent cell, is asexual reproduction. The prefix "a" indicates "without," so asexual reproduction means reproduction without any sexual interaction. Cells that are destined to become **gametes**, egg and sperm cells, will ultimately unite during fertilization to form totally new and unique offspring. This process is called **sexual reproduction**.

Cell Cycle

As each cell lives out its life, it goes through a number of developmental stages. All the stages of a cell's life are called the **cell cycle**. The terms for the major phases of the cell cycle are **interphase** and **mitosis**. Most of an average cell's life is spent in interphase. Interphase is made up of Gap 1, Gap 0, Synthesis, and Gap 2 phases. Like all living things, cells spend a good part of their lives taking care of the day–to-day tasks shared by all living things. Brand new cells spend their time getting nutrients, adding mass, adding and organizing organelles, developing into a mature cell, maintaining homeostasis, getting rid of waste, and just growing up. This growth and maintenance occurs during a stage called Gap One (G_1). At the end of the G_1 stage, the cell has to make a choice. When it reaches the Restriction Point (R), the mature cell can opt to continue as it has been, metabolizing nutrients, doing its job, maintaining itself (G_0) or it can choose to move into a reproductive stage. Several reasons exist for the cell to selecting entering either G_0 or the reproductive cycle: It may have been genetically programmed to live for a certain amount of time before reproducing, it may have received a signal from other cells to reproduce, or it may have lost cell–to-cell contact with surrounding cells.

If the cell enters G_0, it can generally select to move into a reproductive mode whenever that becomes necessary. If the cell passes the Restriction Point, it commits to entering the cellular reproduction cycle. The cell first goes into Synthesis (S) phase. During S phase the chromosomes in the nucleus of the cell get copied. The duplicate chromosome is attached to the original by a **centromere** and together they are called **sister chromatids**. After new chromosomes are produced, the cell needs to make more organelles and material so that there will be enough substance to divide the parent cell into two new, identical daughter cells. This second growth phase is called Gap Two (G_2). Ultimately, at the end of G_2, the parent cell has enough material to enter the final part of the cell cycle, mitosis.

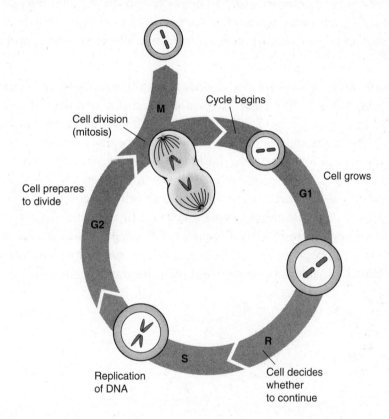

Mitosis

When a cell enters mitosis, it goes through a series of steps that ensures that one parent cell will produce two daughter cells that are identical both to each other and to the original parent cell. This process works to form new and replacement cells; it helps each of our bodies to grow, to maintain themselves, and to heal damage and injuries. Mitosis is made up of four distinct processes that occur in a specific order: Prophase, Metaphase, Anaphase, and Telophase.

1. Prophase is the stage in which chromosome pairs become condensed, clear, and distinct, the nuclear membrane breaks down, the nucleolus disappears, and the centrioles begin to move to opposite poles (ends) of the cell. You can remember the order of prophase in mitosis by recalling that it is the preparatory, preface, or preliminary step; all of these words begin with "p" just like prophase.

2. Metaphase is the phase in which the sister chromatids line up individually along the cells equator. Sister chromatids are held in place by the attachments of their centromeres to the spindle fibers, microtubules anchored at the poles of the cell by centrioles. This alignment prevents both sister chromatids from ending up in the same cell when the cell finally divides. You can remember the order of metaphase in mitosis by recalling that it is the stage when sister chromosomes line up in the middle of the cell; middle and metaphase both begin with "m."

3. During anaphase, the centromeres split and the spindle fibers contract, pulling the chromosomes toward opposite ends of the dividing cell. You can remember the order of anaphase in mitosis by recalling that it is the time when cells move apart or away from each other; apart, away, and anaphase all begin with "a."

4. Telophase is the step when the nuclear membrane begins to reform around the chromosomes, the nucleolus begins to reappear and the one cell begins to separate into two. The dividing process is different in plants and animals. In animal cells, the equator of the cell membrane begins to pinch in; it continues to pinch until it meets in the middle of the cell and finally splits the cell in half. In plants, a cell plate begins to form and ultimately splits the one cell into two. The rigid cell wall prevents the cell from pinching in so the cell plate forms from the middle of the plant cell and finally splits the cell. Telophase ends with **cytokinesis**, the moment when the cell divides. You can remember the order of telophase in mitosis by recalling that it is the phase when the one cell actually divides in two; two and telophase both begin with "t."

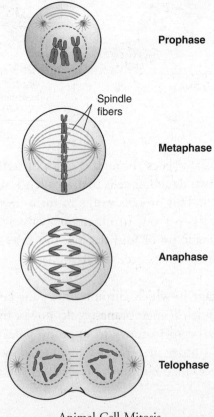

Prophase

Spindle fibers

Metaphase

Anaphase

Telophase

Animal Cell Mitosis

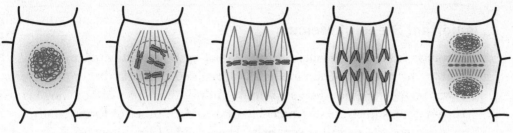

Plant Cell Mitosis

So, mitosis is the way that all living organisms make the cells that they need to grow, develop, conduct metabolism, and to maintain organization and homeostasis. Mitosis allows us to heal after we get a cut, a sunburn, or following surgery. Mitosis lets us have identical replacement cells for aging and damaged **somatic**, or body, cells. Each dividing parent cell is initially **diploid**, meaning that it has two copies of every chromosome. Then during S phase of interphase, every parental chromosome doubles to form matching sister chromatids. As a result, each healthy cell produced during mitosis will have an exact copy of each of their chromosomes. This means that there will be two copies of each chromosome in every new cell, just like their parent cells. The daughter cells will be diploid and each chromosome will carry information for exactly the same traits as the parent and sister cell. The asexual process of mitosis is going on constantly in our bodies and in the bodies of all living things.

Meiosis

The process that allows sexual reproduction to occur is called **meiosis**. Meiosis creates gametes, or reproductive cells, from diploid cells that have two chromosomes for each pair. Gametes are **haploid**, meaning that they only have one chromosome for each pair of chromosomes, so they have to join with another **homologous**, or matching, chromosome to give instructions to make an offspring. The joining of two gametes is called **fertilization**. When one gamete unites with a gamete from another organism of the same species, the different information on each of their paired chromosomes contributes instructions for the assembly of a unique individual, almost always different from its parents and siblings. The reason that the individual is distinctive is that each homologous chromosome has unique information and instructions from a different parent.

The process of meiosis may seem a lot like that of mitosis but there are some very important differences. The end result of mitosis is two complete daughter cells while the end result of meiosis is four haploid gametes, either sperm or egg cells. Ultimately, in order to produce a viable egg, three of these newly formed egg cells become structures called "polar bodies" which contribute almost all of their cytoplasm to the one egg cell from this meiotic division that can become a fertile egg cell. Remember that the goal of mitosis is to create two complete daughter cells, identical to each other and to their parent cell. The goal of meiosis is to create gametes which will each contribute half of the chromosomes necessary to form a totally new and unique individual. The steps of meiosis, in order, are Prophase I, Metaphase I, Anaphase I, Telophase I, Cytokinesis I, Prophase II, Metaphase II, Anaphase II, Telophase II, and Cytokinesis II.

Important Steps of Meiosis

- Prophase I is the stage where chromosomes become condensed and visible, the nuclear membrane breaks apart, and centrioles begin to move toward opposite poles. In a step unlike mitosis, each pair of homologous chromosomes lines up very precisely next to each other to form tetrads, in a pairing process called **synapsis**. These pairs of homologous chromosomes will be the same length and will contain the same genes. However, since one pair of sister chromatids has the mother's genetic information and the other contains the father's genetic information, they will have different instructions for the same genes. During this phase, there is likely to be a recombination of homologous genetic material between chromatids in a process called **crossing over**.

Crossing over during synapsis of prophase I

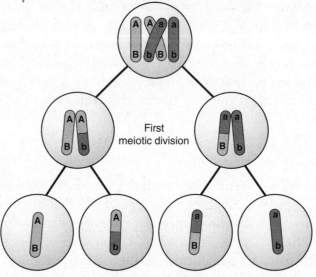

Second meiotic division

- During metaphase I, tetrads, or homologous chromosome pairs, are lined up along the cell's equator. There will be chromosomes from each parent on both sides of the cell's midline.
- Sister chromatids remain attached to each other during anaphase I, but the homologous chromosomes are separated and sent to opposite poles.
- In telophase I, sister chromatids may be contained in a temporary nuclear membrane (though not in all species), and the spindle fibers break apart.
- During cytokinesis I, the cell splits into two cells, each with a full set of chromosomes.
- Prophase II includes the movement of centrioles to opposite poles and spindle fibers reassembling. If a nuclear membrane was formed in telophase I, it now disassembles.
- In metaphase II, sister chromatids are lined up along the cell's equator, or middle.

- During anaphase II, the centromeres split and sister chromatids are pulled apart to opposite poles.
- Telophase II involves the reforming of nuclear membranes around the chromosomes at each pole and the cell begins to split into two cells.
- At cytokinesis II there are now four cells that have come from the original cell. Each new cell contains only one of the four original homologous, duplicated chromosomes, so these cells are called haploid. Each of the four new cells is a potential gamete.

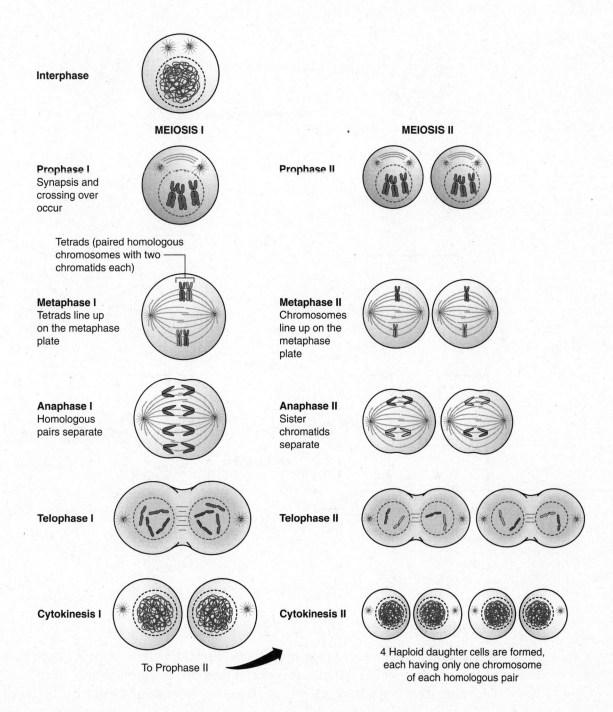

Interphase

MEIOSIS I   **MEIOSIS II**

Prophase I
Synapsis and
crossing over
occur

Prophase II

Tetrads (paired homologous
chromosomes with two
chromatids each)

Metaphase I
Tetrads line up
on the metaphase
plate

Metaphase II
Chromosomes
line up on the
metaphase
plate

Anaphase I
Homologous
pairs separate

Anaphase II
Sister
chromatids
separate

Telophase I

Telophase II

Cytokinesis I

Cytokinesis II

To Prophase II

4 Haploid daughter cells are formed,
each having only one chromosome
of each homologous pair

Haploid cells that that are produced during meiosis must still undergo **gametogenesis** before becoming viable, or usable, egg or sperm cells. A sperm cell must compact its DNA and eliminate almost all of its organelles and cytoplasm, until it's essentially a cluster of chromosomes surrounded by a membrane, attached via a mitochondria-(energy) packed segment to a powerful flagellum. In the production of an egg cell, only one of the four haploid cells produced by meiosis will become an egg, the other three become polar bodies. The cell that will become an egg cell receives most of the cytoplasm and organelles during meiosis; this unequal distribution will provide a fertilized egg with the nutrients and molecules it will need to grow.

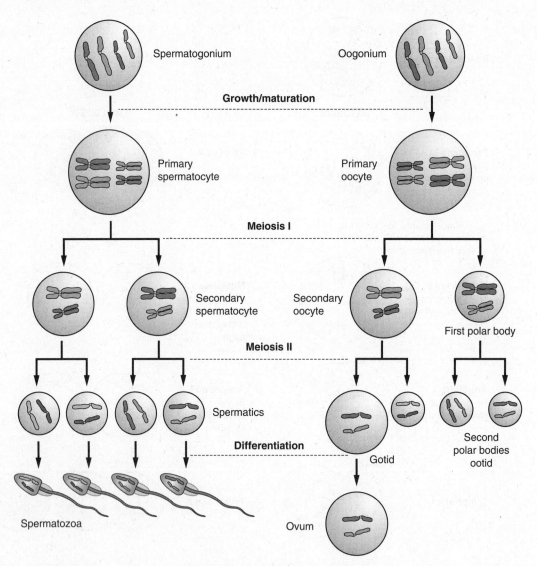

Gametogenesis: Sperm and Egg

Telomeres

A **telomere** is a sequence of repeating DNA nucleotides that is found on both ends of each chromosome. Telomeres are the end caps on our chromosomes. They have been described as similar to the plastic wraps on the ends of a shoelace. Their job is to control cell reproduction. When telomeres are working correctly they control how many times a cell reproduces. The enzyme telomerase works to add repeating DNA sequences to the ends of chromosomes to maintain the telomeres. Telomerase becomes less active as we age and is inactive in most adult cells; it is active, though, in stem cells, gamete-producing cells, hair follicles, and in most cancer cells.

The majority of any organism's cells have telomeres that are reduced by each episode of cellular reproduction, so there are a limited number of times that most cells can divide. As the telomere gets smaller, there is more likelihood of mistakes being made in cellular reproduction. Ultimately, when the telomere is gone, the cell can no longer reproduce and will die. The aging of cells, **senescence**, is the result of reduced telomeres. As telomeres get smaller, they provide less protection against mistakes in the copying of genetic information. Have you ever noticed that a baby usually has smooth, consistent skin and that elderly people often have irregular skin with spots and bumps? Well, the baby has longer telomeres that better protect them from mistakes and variations in asexual reproduction of the skin cells. As a baby's skin cells age, the parent cells produce daughter cells that are identical to the parent and to sister cells. As an elderly person's skin cells age, they don't have the strong telomeres to control their reproduction so daughter cells may be different from their parents and from each other. This causes variations in the skin like age spots and skin tags.

Apoptosis is programmed cell death and is a result of degraded telomeres. When the telomere is gone, or almost gone, from the end of an organism's chromosomes, the cells containing those chromosomes will die. If the life expectancy of a dog breed is 11 to 13 years, a mouse is 2 to 3 years, or a Galapagos tortoise is more than 200 years, it is because their cells are programmed genetically to undergo apoptosis at about that time. If telomeres do not degrade, or if they continue to be rebuilt by telomerase, cells can live for much longer than their normal life expectancy. This is called cellular **immortality**. When cells become immortal they stop following the rules for reproduction and this causes cancer. Cancer is a situation where cells have gone wild; they overproduce and they keep reproducing even when they're crowded, causing tumors. Immortal cells will begin invading tissues where they don't belong. There is a lot of scientific research happening to better manipulate and understand telomeres and telomerase. The hope is that we will gain a better understanding and control of the processes of aging and cancer.

Important Terms to Remember

Apoptosis: Programmed cell death; the result of telomere destruction.

Asexual reproduction: The process of making identical, duplicate cells from one original parent. Somatic cells reproduce asexually.

Cell cycle: All the stages of a cells life; interphase and mitosis.

Centromere: Structure that holds sister chromatids together and which attaches sister chromatids to spindle fibers during mitosis.

Crossing over: During Prophase I of meiosis, homologous sister chromatids align with each other and segments of DNA are exchanged. This is a major reason for variation between family members and members of a species.

Cytokinesis: Cell division; the point at which a parent cell undergoing mitosis becomes two daughter cells, each identical to one other and to the parent cell.

Diploid: Cells that contain two homologous copies of each chromosome; each chromosome will have potentially different information for the same genes.

Fertilization: The uniting of egg and sperm cells producing a single new organism with genetic information from two parents.

Gametes: Sex cells; egg and sperm.

Gametogenesis: The formation of haploid gametes.

Haploid: Cells that contain one copy of each chromosome; haploid cells are generally gametes.

Homologous: A matching chromosome; it has the same genes but may have different information for those genes; for example, eye color or hair color are on the same genes on homologous chromosomes but may code for different variations of that characteristic.

Immortality: Cellular immortality results in uncontrolled mitosis. Immortal cells stop following the rules for reproduction and can live for much longer than they should. Uncontrolled mitosis can result in cancer.

Interphase: The segment of the Cell Cycle in which cells grow, develop, metabolize nutrients, maintain organization, and homeostasis. If a cell has committed to reproducing, DNA will be copied during S phase of interphase. Interphase is made up of the G_1, G_0, S, and G_2 phases.

Meiosis: Process of making haploid gametes; steps are Prophase I, Metaphase I, Anaphase I, Telophase I, Cytokinesis, Prophase II, Metaphase II, Anaphase II, and Telophase II.

Mitosis: Process of making two daughter cells each identical to one another and to their parent cell; steps are Prophase, Metaphase, Anaphase, and Telophase.

Senescence: Aging of cells; as telomeres shorten, cellular reproduction becomes less precise and more mistakes happen in the production of replacement cells.

Sexual reproduction: Egg and sperm cells unite during fertilization to form totally new and unique offspring.

Sister chromatids: When chromosomes are duplicated during S phase, duplicate chromosomes are attached to each other by a centromere; this helps them to align correctly during metaphase.

Somatic Cells: Body cells, all cells that are not gametes.

Synapsis: The pairing of homologous chromosomes during prophase I of meiosis; when homologous sister chromatids exchange DNA, it allows variation through genetic recombination.

Telomere: A sequence of repeating DNA nucleotides that form end caps at both ends of every chromosome. Telomeres minimize the breakdown of DNA as a result of repeated copying and protect the integrity of cellular reproduction.

Practice Questions

Multiple-Choice

1. You are looking through a microscope at a slide with a piece of onion. You can see an elongated cell that has two nuclei, one at each end, and you see a line forming in the middle of the cell. Your teacher tells you this "dividing line" is called a _____.

   (A) plasma membrane
   (B) spindle apparatus
   (C) mutation
   (D) cell plate

2. You are still looking at the same onion specimen and you see two small cells that together take up little more space than most single cells. What has probably happened recently to this cell?

   (A) meiosis
   (B) synapsis
   (C) cytokinesis
   (D) senescence

3. The two young cells you were just looking at are now likely to be in which phase of the cell cycle?

   (A) mitosis
   (B) G_1 phase
   (C) G_2 phase
   (D) meiosis

4. You've drawn the following image of what you observed through your microscope. What are you looking at?

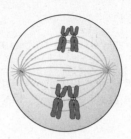

(A) mitosis, metaphase
(B) mitosis, anaphase
(C) meiosis, metaphase I
(D) meiosis, anaphase I

5. You've drawn the following image of what you observed through your microscope. What are you looking at?

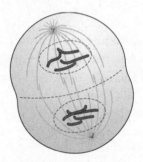

(A) mitosis, prophase
(B) mitosis, telophase
(C) meiosis, prophase I
(D) meiosis, telophase I

6. You've drawn the following image of what you observed through your microscope. What are you looking at?

(A) mitosis, prophase
(B) meiosis, prophase I
(C) mitosis, anaphase
(D) meiosis, anaphase I

7. You've drawn the following image of what you observed through your microscope. What are you looking at?

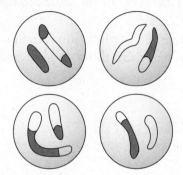

(A) mitosis, telophase
(B) meiosis, telophase I
(C) meiosis, prophase II
(D) meiosis, telophase II

8. A cell has just passed the restriction point and is going to begin the process of cellular reproduction. What is the first thing the cell needs to do?

(A) Make more cytoplasm and organelles.
(B) Copy its DNA.
(C) Enter mitosis.
(D) Assemble its centrioles and spindle fibers.

9. What stage is the cell in question #8 entering?

(A) G_1
(B) G_2
(C) S
(D) G_0

10. When a cell is first formed during cytokinesis, the first stage that it enters is
_____.

(A) mitosis, prophase.
(B) meiosis, prophase I
(C) interphase, G_0
(D) interphase, G_1

Free-Response

1. Hutchinson-Gilford progeria is a genetic disease in which aging happens very quickly and causes young children have the appearance of very old people. From what you've learned in this chapter, what do you think might be happening to progeria patients?

2. There is a couple on television with 18 children. None of the siblings are identical to each other or to their parents. How would you explain this variety of appearances among members of the same family?

3. Some cancer research examines cellular immortality. We often hear about living forever, immortality, as a positive thing. Why is cellular immortality viewed as a negative?

Answer Explanations

Multiple-Choice

1. **D** When plant cells undergo mitosis, they form a cell plate to divide the two new cells.

2. **C** These onion cells have recently undergone cytokinesis. They were just formed when their parent cell underwent mitosis. They are not reproductive cells and so would not be involved in meiosis or synapsis and are not likely to be very old, so they won't be senescent.

3. **B** The two new cells are most likely in G_1, they need to grow and develop to be the same size as the nearby mature cells. These cells are not big, or mature, compared to nearby cells so they are not preparing to reproduce as they would be in G_2, mitosis, or meiosis.

4. **C** You see homologous sister chromatids lined up along the equator of the cell; it has to be meiosis and it has to be metaphase I.

5. **B** This is a mitotic cell in telophase. You can tell it is mitotic because it has separate, homologous chromosomes and you can tell it's in telophase because the nuclei are reforming and the cell is pinching off in preparation for dividing.

6. **B** This cell is undergoing meiosis, it contains aligned pairs of homologous sister chromatids. You know it's in prophase because the centrioles are together and the nuclear membrane is present but breaking apart.

7. **D** This image is clearly the final result of meiosis. It is from the end of telophase II. Mitosis only yields two daughter cells. Each of the resulting cells has genetically recombined individual chromosomes.

8. **B** The first thing the cell does after leaving G_1 is to copy its DNA. After that it makes more cytoplasm and organelles, enters mitosis, and begins to assemble its centrioles and spindle fibers.

9. **C** This cell is entering S phase, the only phase when a cell copies its DNA.

10. **D** Immediately following mitosis and cytokinesis, the new cell enters G_1 of interphase. It doesn't enter G_0 until it's grown up and chosen to not yet become reproductive.

Free-Response Suggestions

1. **Hutchinson-Gilford progeria is a genetic disease in which aging happens very quickly and causes young children have the appearance of very old people. From what you've learned in this chapter, what do you think might be happening to progeria patients?**

   Progeria patients may have a problem with their telomeres that causes premature senescence. If their telomeres are small to begin with, if they break down more quickly than average, or if they never get fully assembled because the telomerase doesn't work right, then DNA will get copied with more mistakes than it should for a healthy child.

2. **There is a couple on television with 18 children. None of the siblings are identical to each other or to their parents. How would you explain this variety of appearances among members of the same family?**

   During meiosis, synapsis and crossing over contribute to variation within a species.

3. **Some cancer research examines cellular immortality. We often hear about living forever, immortality, as a positive thing. Why is cellular immortality viewed as a negative?**

   When cells develop immortality, they are no longer following the life plans set out in their DNA. These cells no longer show cell-to-cell contact inhibition, so they keep reproducing even when they're already crowded. They don't die when they stop functioning correctly.

Molecular Genetics

CHAPTER 6

Outline for Organizing Your Notes

I. Nucleic Acids
   A. DNA
      1. Nucleotide: Deoxyribose + Phosphate Group + Nitrogen Base
      2. Nitrogen bases: Adenine + Thymine and Cytosine + Guanine
      3. Double Helix
      4. Code Held in the Sequence of Nitrogen Bases
      5. Stays Inside Nucleus (exceptions in a later chapter)
   B. Chromosomes (46 in humans: 22 pairs of autosomes and 1 pair of sex chromosomes)
      1. Stay in Nucleus
      2. Made Up of Genes
         a. Genes Are Sequences of DNA That Give Instructions for the Assembly or Use of a Protein
      3. Autosomal Chromosomes Are Diploid (2n) Have Chromosomes in Pairs
      4. Sex Chromosomes Are Haploid (n) Have Single Chromosomes
   C. RNA
      1. Nucleotide: Ribose + Phosphate Group + Nitrogen Base
      2. Nitrogen Bases: Adenine + Uracil and Cytosine + Guanine
      3. Single Strand
      4. Code Held in Sequence of Nitrogen Bases
      5. Made Inside Nucleus, Travels Out Through Nuclear Pore, Enters Cytoplasm

      6. Three types of RNA:
         a. mRNA (messenger RNA) Takes Information from DNA to Ribosome
         b. rRNA (ribosomal RNA) Makes Up Ribosomes; Sites of Protein Synthesis
         c. tRNA (transfer RNA) Brings Amino Acids to Be Assembled in Correct Order

II. DNA Synthesis
   A. Cell Cycle
      1. Life Span of a Cell (G_1, G_0, S, G_2, mitosis)
      2. DNA is Copied During S Phase
   B. DNA Replication
      1. Enzymes Unwind and Unzip DNA Double Helix at Multiple Locations
      2. Enzymes Attach Free Nucleotides to Matching and Available Nitrogen Bases
      3. Original Strands of DNA (templates) Are Now Matched with Complementary Strands
      4. Two Strands of DNA Where There Was One
      5. Cell Now Has Enough DNA to Make Two New Copies of Itself

III. Transcription
   A. Why Transcribe?
      1. To Make Structural Proteins
      2. To Make Functional Proteins

B. Making RNA
1. Protein Needed; Transcription Complex (TC) Is Formed
2. TC Unwinds and Separates Nitrogen Bases at the Gene for Chosen Protein
3. Complementary Nucleotides Are Assembled to Form mRNA
4. The mRNA Leaves the Nucleus Through Nuclear Pore
5. DNA Strand Closes and Re-coils

IV. Translation
A. Reading RNA
1. Ribosome Reads mRNA in Groups of Three Nitrogen Bases Called Codons
2. Each Codon Is Associated with One of the 20 Amino Acids That Make Living Things
B. Making Proteins
1. The mRNA Moves Along a Ribosome
2. The Ribosome Reads mRNA Codons
3. The Ribosome Signals to tRNA with Anticodon That Is Complementary to Current Codon

4. The Codon and Anticodon Bond Temporarily; At Other End of tRNA Is Matching Amino Acid
5. The mRNA Shifts Down One Codon and the Next Complementary tRNA Attaches to It
6. Amino Acids at the Other End of tRNA Form Peptide Bonds in Their Order of Attachment
7. The tRNA and Its Anticodon Fall Away
8. As mRNA Moves Along Ribosome, 100 to 1,000s of Amino Acids Bond Forming a Protein

V. History
A. Early Discoveries (1928 to approximately 1950)
1. Frederick Griffith
2. Oswald Avery
3. Alfred Hershey and Martha Chase
B. Understanding DNA (1950–1958)
1. Erwin Chargaff
2. Linus Pauling
3. Rosalind Franklin and Maurice Wilkins
4. James Watson and Francis Crick

So, What's All the Fuss About DNA…and Why Do We Need to Know About It Anyway?

Think about yourself, your pet (dog, cat, hamster, iguana), a tulip growing in front of your house, the mildew growing in the shower that you were supposed to clean, and the bacteria living in the yogurt in your refrigerator. How are they connected to each other? Living organisms have a relationship to all other living organisms. We share characteristics such as growth, development, use of energy, disposal of waste, homeostasis, and organization. All living things are either cells or are made of cells. Beyond this, though, is an even deeper connection. Living organisms reproduce themselves. This isn't just about producing offspring. Living cells are continuously being repaired and replaced. When a Galapagos tortoise dies at the age of 160, very few (if any) of the cells that exist in its body were present at the time of its birth. Our bodies have the plans in place to make, what should be, exact copies of the cells with which we were born. It is usually to our advantage to replace our old or damaged cells with exact copies of the originals. Our body's instructions for the production and assembly of new and replacement cells, and for the production of offspring, is found in our DNA and our RNA.

Nucleic Acid

One of the four major types of organic molecule is the **nucleic acid**. Nucleic acids are either **DNA** or **RNA**. DNA stands for deoxyribonucleic acid and RNA stands for ribonucleic acid. Nucleic acids store information for the parts and assembly of every living thing on Earth. This information is stored in the order, or sequence, of **nitrogen bases** that make up the nucleic acids. Nitrogen bases are held in a very specific order by anchors made up of sugars attached to each other using phosphate groups. The sugars that hold nitrogen bases in order are <u>d</u>eoxyribose in <u>D</u>NA and <u>r</u>ibose in <u>R</u>NA. Nucleic acids are large molecules that are made up of many smaller subunits called **nucleotides**. Nucleotides are relatively simple molecules made up of three parts; a sugar (deoxyribose or ribose) bonded to a phosphate group on the top or bottom and to a nitrogen base off to the side.

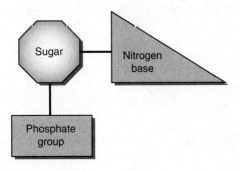

An Analogy

Let's consider the steps in building a house. First, a plan needs to be made. This plan will include a parts list and directions for putting those parts together; usually this will be in the form of a blueprint. A blueprint will include every item that should be used, its size, what it's made of, where and how it should be positioned, and how it should interact with surrounding items. The cells, tissues, organs, organ systems and, ultimately, the bodies of all living things are assembled with the same type of a plan. This plan isn't in the form of a paper blueprint though, the plan is found in our Nucleic Acids.

DNA

Deoxyribonucleic acid is made up of millions of nucleotides attached together. They're attached in a precise order that looks like a spiral ladder, called a **double helix**. The sides of the ladder are made of Sugar (Deoxyribose)—Phosphate—Sugar—Phosphate—Sugar—Phosphate…repeated over and over. The steps, or rungs, of the ladder are made using the **nitrogen base** pairs. These are the only place where variation occurs. There are four nitrogen bases in DNA; Adenine (A),

Cytosine (C), Guanine (G), and Thymine (T). The order of these nitrogen bases gives our existing cells the instructions to build new cells.

Based on their molecular structure, adenine and guanine (with larger double carbon-ring bases) are called **purines** and cytosine and thymine (with smaller single-ring bases) are **pyrimidines**. Nitrogen bases are paired very specifically. In DNA, adenine only bonds with thymine and the bond that holds them together is a double hydrogen bond. Cytosine and guanine bond uniquely to each other with a triple hydrogen bond.

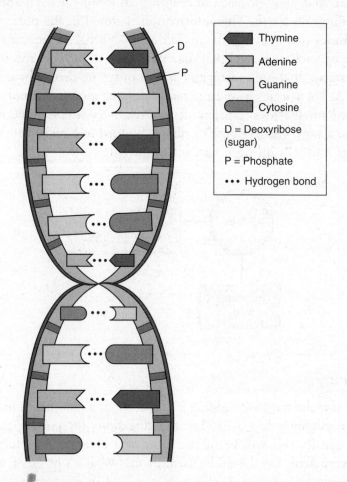

The sequence of nitrogen bases in DNA is often represented like this:

ATTATATAAAGCCACGCATAC
TAATATATTTCGGTGCGTATG

As you can see, A + T and C + G are always paired in DNA. These 2 matching DNA strands are called **complementary**. It is in this sequencing and base pairing that information is stored for the accurate formation and assembly of new, living cells and organisms.

Back to That Analogy

So, let's revisit our house building analogy. The DNA is like a blueprint for a house. The paper and the ink of the blueprint are like the sugar—phosphate backbone. The paper and ink don't change and they don't alter the information that they carry, they simply allow information to be organized, made intelligible, and readable. The blueprint will have symbols that represent all the parts that make up the house: the wood, the plumbing, the wiring, the beams, and the fixtures. The blueprint will also include symbols to represent where these parts will be placed and will include codes that represent the rules to be followed in building the house. All these symbols and codes are like the nitrogen bases, every bit of building information is carried in them. They need to be written correctly, read correctly, interpreted correctly, and the instructions carried out in order to build the house that you intend.

Chromosomes

DNA is primarily found in the nucleus of a eukaryotic cell. In prokaryotes, DNA is found floating in the bacterial cytoplasm. DNA is the material that makes up a **chromosome**. Within each chromosome are segments called **genes**, each with anywhere from 100 to more than two million base pairs. Each gene directs the formation, activation, function, or regulation of a product that is necessary to the healthy development of a living organism. Different types of organisms have different numbers of chromosomes. Organisms that reproduce sexually carry two copies of each chromosome in their nuclei. In these organisms, cells that have two copies of each chromosome are called **diploid** and cells with a single copy of each chromosome are called **haploid**. Body cells, or **autosomes**, are diploid and, in humans, contain 46 chromosomes. Sex cells, eggs and sperm, are haploid. Sex cells are known as **gametes** and, in humans, contain 23 chromosomes. Different species have different diploid numbers; camels have a diploid number of 70 (and a haploid number of 35), mosquitoes have a diploid number of six, and a tomato plant has a diploid number of 24.

RNA

Ribonucleic Acid is single stranded (unlike double-stranded DNA) and is also made of nucleotides. RNA looks like one side of the ladder that makes up DNA. It has a single side rail made up of alternating sugar (ribose, instead of deoxyribose) and phosphate groups. Attached to each ribose molecule in RNA is a nitrogen base. These bases include adenine (A), cytosine (C), and guanine (G). RNA does not contain thymine (T) and instead uses a similar molecule called uracil (U).

There are three types of RNA that perform three different functions. Messenger RNA (mRNA) is a strand of RNA that is a copy of the DNA instructions and that can carry those plans out of the nucleus into the cells cytoplasm to be copied at the ribosome. Ribosomal RNA (rRNA) makes up a large part of each ribosome. Ribosomes, you'll remember, are the organelles that are responsible for assembling proteins for our cells. Transfer RNA (tRNA) is a structure that carries specific amino acids through the cytoplasm to the ribosome where the amino acids can be assembled into the proteins listed in the DNA instructions.

DNA Replication

Cell Cycle

Different cells in every organism have different life spans. Depending on the type of autosomal (non-reproductive) cell, it may be programmed to live for days, weeks, months, or years. The cell's lifetime is called the **cell cycle**. The beginning of the cell cycle, a stage called G_1, is the time of growth and development. It begins when the cell is first formed and ends at the point where the cell either enters G_0 or S Phase. If the cell is going to maintain itself without focusing on copying or reproducing itself, it enters G_0; taking in nutrients, eliminating waste, and doing its job. If the cell is going to reproduce itself, it moves past G_0 through the Restriction Point and into S Phase. During S Phase of the cell cycle, DNA is synthesized, or copied, in preparation for mitosis.

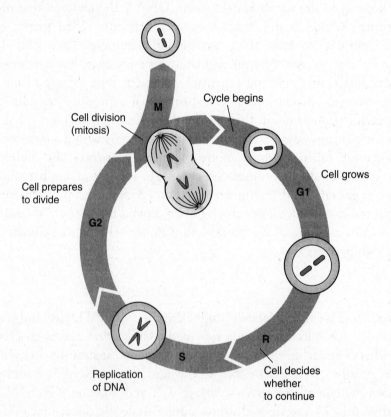

DNA Replication

DNA is copied by opening up the double helix between the nitrogen bases that are paired together. The double hydrogen bonds between A+T and the triple hydrogen bonds between C+G are broken and the DNA strand opens like a zipper that has split apart even though it is still zipped. Enzymes, proteins that trigger cellular reactions, open the original DNA strand at many locations, **origins of replication**, creating **replication bubbles**. Other enzymes hold the DNA apart. Still other enzymes help nucleotides that are free floating in the nucleus, to attach to both of the two original strands of the DNA. Ultimately, both strands of the one original DNA helix are copied. The two original strands of the DNA double helix are called templates; this means that they are patterns that are used to produce complementary DNA strands. This process is considered **semiconservative** because one strand is kept, or conserved, from the original DNA while the other strand is newly formed. When DNA synthesis, or replication, is complete the cell has two identical copies of the one original DNA strand.

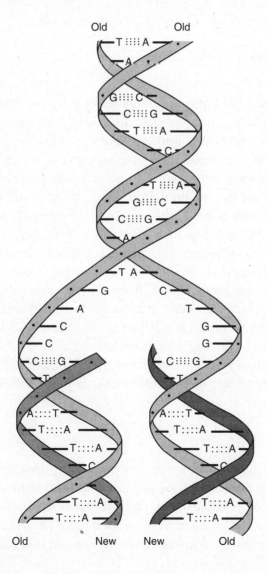

After all of the cell's chromosomes have been duplicated, the cell moves into the G$_2$ phase of the cell cycle. During this time the cell forms more cytoplasm and organelles so that when it finally enters mitosis (or meiosis), it will have all the parts it needs to become two (or four) new cells. Remember, this process is going on all the time in thousands of the cells in your body, your pet chinchilla's body, and in the oak tree at the park.

That Analogy Again!

We started with our blueprint in the form of an original strand of DNA. The job of DNA replication is to make additional copies of the DNA, so that during synthesis, we've made more copies of our blueprint in order that more houses can be built using the same specifications as the original house.

Transcription

Why Transcribe?

DNA has more jobs than just copying itself so that new cells can be formed. DNA is also responsible for directing the assembly of proteins. Proteins are important because other than water, proteins make up the highest percentage of material in the bodies of living organisms. Bodies require proteins for growth and development, for repair and maintenance, for replacement of damaged or old cells and tissues, to act as digestive and metabolic enzymes, and to serve as an ingredient in hormones. Proteins can be structural proteins like those in actin and myosin, which allow contraction of our muscle cells, or like fibrillin, which acts as a type of glue that holds some of our cells together. They can be functional proteins, like insulin or Hexosaminidase-A, an enzyme that prevents a specific type of fat from building up in our brains. If actin or myosin are not assembled correctly, they can cause muscular problems and heart disease. If fibrillin is made incorrectly it can cause Marfan syndrome, which can result in retinal detachment and that has the potential to cause a fatal aortic dissection. When insulin doesn't work correctly the result is diabetes. Incorrectly formed Hex-A causes the accumulation of lipids in the brains of babies with Tay-Sachs disease and these children will go blind, deaf, and will die by around the age of five. Obviously, copying DNA correctly is a very important job.

Making RNA

In order to make a protein, the whole strand of DNA does not need to be copied, only the required gene needs to be copied. The process of copying a gene into the language of RNA is called **transcription**. Transcription begins when the cell identifies a need for the production of some type of protein. Enzymes and other proteins combine to form a transcription complex and this complex will recognize the beginning of the gene that codes for the necessary material. The transcription complex starts unwinding and separating the gene's segment of DNA. One side of the DNA strand serves as a template while complementary RNA nucleotides bond, temporarily, with the gene's nitrogen bases. The newly-formed RNA strand hangs from the DNA while more RNA nucleotides are added. DNA that has already served as a template closes back up and will be available as a template the next time this protein needs to be made. The new molecule, a strand of mRNA, ultimately leaves the DNA, exits the nucleus through a nuclear pore and approaches the ribosome. It moves along and is interpreted by the ribosome; it has served its function as a messenger.

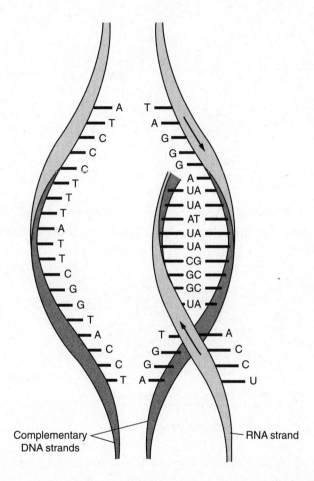

Complementary DNA strands — RNA strand

As you know, the completed new segment of RNA is different from DNA in several ways. The RNA is made with the sugar ribose instead of deoxyribose. RNA is single stranded, instead of double stranded like DNA. Thymine is not found in RNA; instead RNA uses uracil to pair with adenine. Also, RNA is able to leave the nucleus of the eukaryotic cell while DNA remains inside the nucleus. Finally, DNA gives instructions to make three types of RNA found in eukaryotic cells:

- Messenger RNA (mRNA) is the molecule that copies the nitrogen base sequence of the gene and turns the DNA language into RNA language, a code that can be understood and interpreted in the cytoplasm.

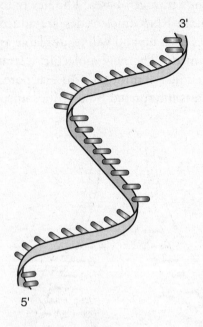

- Ribosomal RNA (rRNA) is the form of RNA that makes up a large part of the ribosomes, the cellular organelles where proteins are assembled.

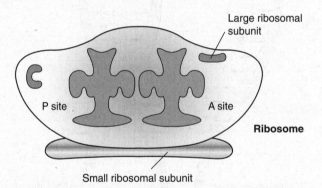

- Transfer RNA (tRNA) is a form of RNA that is responsible for bringing the correct amino acids to the ribosome to be assembled in the correct order to make the needed proteins.

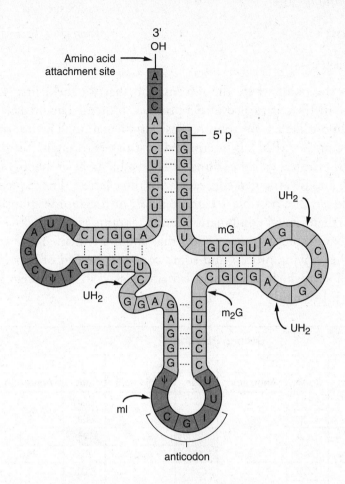

A quick and easy comparison of DNA and RNA basics:

DNA	vs.	**RNA**
Deoxyribonucleic Acid		Ribonucleic Acid
Double Stranded		Single stranded
Deoxyribose Sugar		Ribose Sugar
Thymine bonds to Adenine		Uracil bonds to.Adenine
Generally found in the nucleus		Found in nucleus and cytoplasm

The Analogy

The general foreman and construction supervisor are like our molecules of mRNA. They take the blueprint, read it, and convert (or transcribe) it from symbols and codes into a language that is understandable. They bring this information to the construction team at the building site, the ribosome. The job site with its construction crew is similar to the rRNA. The tRNA parallels the delivery people and their trucks, bringing the correct supplies to the place where they'll be assembled into the final product, a house.

Translation

Reading RNA

Translation is the point where the information that was held first in the code of DNA, then in mRNA, is turned into a product, protein. The mRNA travels along a ribosome almost like a train running past a train station. The information in the nitrogen bases of the mRNA is interpreted as it passes through. Nitrogen bases are read in groups of three, called **codons**. It is possible to assemble the four nitrogen bases of RNA into 64 arrangements each with three letters. The sequence of codons represents one out of a possible 20 amino acids, or it can order translation to start or stop. Since there are 64 combinations and 20 amino acids, it's obvious that most amino acids have more than one codon that codes for them. Some codons trigger the start of the copying process and some code for the end of copying. The table below can tell you which amino acid is being placed in the protein and where in the sequence it belongs.

		Second Position								
		U		**C**		**A**		**G**		
		Code	Amio acid	Code	Amio acid	Code	Amio acid	Code	Amio acid	
First Position	**U**	UUU	phe	UCU	ser	UAU	tyr	UGU	cys	U
		UUC		UCC		UAC		UGC		C
		UUA	leu	UCA		UAA	STOP	UGA	STOP	A
		UUG		UCG		UAG	STOP	UGG	trp	G
	C	CUU	leu	CCU	pro	CAU	his	CGU	arg	U
		CUC		CCC		CAC		CGC		C
		CUA		CCA		CAA	gin	CGA		A
		CUG		CCG		CAG		CGG		G
	A	AUU	ile	ACU	thr	AAU	asn	AGU	ser	U
		AUC		ACC		AAC		AGC		C
		AUA		ACA		AAA	lys	AGA	arg	A
		AUG	met	ACG		AAG		AGG		G
	G	GUU	val	GCU	ala	GAU	asp	GGU	gly	U
		GUC		GCC		GAC		GGC		C
		GUA		GCA		GAA	glu	GGA		A
		GUG		GCG		GAG		GGG		G

(Third Position column: U, C, A, G repeated for each First Position block)

To determine which amino acid is represented by a codon, look at the column on your far left. It says "first position UCAG." Any codon that begins with a U will be found across the top of the chart and any that begins with a G will be at the bottom. A bar across the top of the table is labeled "second position UCAG." If the second letter of the codon is a C or an A, it will be found in the middle of the table somewhere between the top and bottom. The third position is listed on your far right; UCAG is repeated once for each of the first position letters. So, for example, if you want to identify the amino acid coded for by the RNA nitrogen bases CCG, you'll look at the first position C; as you read the table from left to right, you'll see the amino acid abbreviations LEU (leucine), PRO (proline), HIS (histidine), GLN (glycine), and ARG (arginine). Next, you look at the column in the table under the second position C. Although there are four different possibilities—SER (serine), PRO, THR (threonine), and ALA (alanine)—you know that the amino acid coded for by CCG will be PRO because that's the only one that has C in the first position. In this case the third position won't make a difference because, this time, all four nitrogen bases in the third position code for the same amino acid, proline.

Making Proteins

So, we left the mRNA rumbling though the ribosome like a train going through a station. As it goes through, the first codon is read and attracts a complementary tRNA molecule. This tRNA molecule has two parts that are important for translation; at one end is an **anticodon** and at the other end a matching amino acid. The anticodon is a sequence of nitrogen bases that is complementary to the codon being read by the ribosome; for example if the codon were AUG the anticodon would be UAC. At the other end of this same molecule of tRNA, would be a molecule (you can find it on the table on page 94) of the amino acid, MET (methionine). The anticodon bonds for a short time with the codon, the mRNA moves along and the next codon attracts its anticodon, tRNA, and amino acid. The methionine forms a peptide bond with the next amino acid and that forms a bond with the next amino acid, and so on until the stop codon is reached. As the peptide bonds are formed and protein is assembled, the tRNA drops off and goes back to find another copy of its amino acid in the cytoplasm. The stop codon indicates that production of this protein is complete.

> ### The Last DNA Analogy
>
> Translation in an organism represents the assembly and construction of all the parts that make up the house. Each protein that's assembled is one of the structural or functional components of the completed house. As long as an organism is alive, transcription, translation and protein synthesis continue to occur. The house is maintained, modified, and renovated.

History of Understanding Molecular Genetics

Early Discoveries

The scientific process that resulted in our current understanding of molecular genetics began less than 100 years ago. In 1928, a microbiologist named Frederick Griffith identified a material that could change a harmless form of bacteria into a disease causing bacteria. He called this material the "transforming principle." In 1944, a biologist, Oswald Avery, and his team developed a process that allowed them to observe this transformation. They then found a way to purify the material and identify that the transforming principle was DNA. Many scientists thought that DNA was too simple and repetitive to be the transforming factor and that the material had to be a protein. The evidence that confirmed that DNA was the molecule that carries genetic information came from Alfred Hershey and Martha Chase in 1952. Hershey and Chase infected bacteria with viruses that had been tagged with radioactive phosphorus. Because protein does not have phosphorus and DNA does (remember, it's found in the sugar–phosphate backbone of the DNA strand), the presence of radioactivity inside the infected bacteria indicated that DNA, in fact, was the material that holds the genetic code.

Understanding DNA

It was important that scientists knew that DNA carried genetic information but, by the early 1950s, they still didn't understand the form of the genetic code or how this information could be stored. In 1950, Erwin Chargaff identified the relationship between adenine/thymine and cytosine/guanine when he established that in DNA, the amount of adenine roughly equals the amount of thymine, and cytosine approximately equals guanine. At around the same time biochemist Linus Pauling identified the spiral, or helix, shape of some proteins. In another lab, Rosalind Franklin and Maurice Wilkins used X-ray crystallography to photograph DNA; they saw that DNA's structure was an X surrounded by a circle. Wilkins shared this information with James Watson and Francis Crick without Franklin's permission. In 1953, Watson and Crick published a paper explaining the double helix model they had developed for DNA. Francis Crick used all this information to publish his *Central Dogma of Molecular Biology* in 1958, in which he described the progression of hereditary information from DNA to RNA to proteins. There has been a lot of new information about genetics in the last 50 years but the *Central Dogma of Molecular Biology* has endured as the foundation for our understanding of heredity.

Important Terms to Remember

Anticodon: Sequence of three nitrogen bases carried by tRNA; this sequence is complementary to an mRNA codon and serves to hold the correct amino acid in place to be assembled into the protein directed by the genetic code.

Autosome: Chromosomes that are not directly related to the sex of an organism.

Cell cycle: The life time of a eukaryotic cell; includes growth, development, metabolism, DNA synthesis, and cell division.

Chromosome: Strand of DNA found in a cell's nucleus; it contains a series of genes and regulatory information (which guides the functions of manufactured proteins).

Codon: Sequence of three nitrogen bases in a strand of mRNA; these determine which amino acid is required and where it's to be placed in the protein that's being made.

Complementary: One strand of the DNA molecule or the strand of RNA that matches the original DNA strand; they are not identical because the original, or template, side of the strand will contain adenine and the complementary strand will have either thymine (in DNA) or uracil (in RNA) or it will contain cytosine complemented by guanine, or thymine will have adenine, or guanine will have cytosine. Anticodons are also complementary to the codons of mRNA.

Diploid: Organisms that reproduce sexually have two copies of each chromosome in their autosomal, or body, cells; this is abbreviated 2n.

DNA: Deoxyribonucleic Acid; the molecule that holds genetic information for all living things.

DNA synthesis: Process that produces an exact copy of each original strand of DNA.

Gamete: Sex cell in all sexually reproductive species; egg or sperm.

Gene: Segment of a chromosome that contains the code for a particular protein.

Haploid: Organisms that reproduce sexually have one copy of each chromosome in their gametes, or sex cells; this is abbreviated n.

Nitrogen base: Component of nucleotide (along with sugar and phosphate group) that directs the specific sequence of genetic code. For example, the nitrogen bases ACC in DNA, code for the mRNA codon UGG, which codes for the anticodon ACC, which brings along the amino acid tryptophan to be linked in order into a new protein.

Nucleic Acid: Organic macromolecule containing genetic information; a polymer of nucleotides.

Nucleotide: Monomer of nucleic acids; made of sugar + phosphate group + nitrogen base.

Protein synthesis: End result of transcription and translation; the protein product is what guides the construction of, and provides the materials for, the assembly of all living things.

Purine: Large nitrogen bases with double carbon-ring structures; adenine or guanine.

Pyrimidine: Small nitrogen bases with single carbon-ring structures; uracil, cytosine or thymine.

Replication: Making an exact copy of the original, for example when DNA needs to be copied during S Phase of Interphase in order to have enough DNA to undergo mitosis.

RNA: Ribonucleic Acid; the molecule that carries genetic information to guide formation of proteins.

Semiconservative: When DNA is copied the result is two new strands each having one original strand and one new, complementary strand; this process saves and uses the original strand.

Transcription: Process of using the nitrogen base sequence in DNA to build a strand of RNA.

Translation: Process of using the nitrogen base sequence in mRNA to assemble a protein molecule.

Practice Questions

Multiple-Choice

1. Which DNA strand is complementary to the following segment of DNA?

   AATATACGTACCGCG

   (A) CCGCGCATGCAATAT
   (B) AATATACGTACCGCG
   (C) TTATATGCATGGCGC
   (D) UUAUAUGCAUGGCGC

2. In the DNA strand seen in Question #1, how many amino acids will be added as a result of this sequence?

   (A) 2
   (B) 15
   (C) 1
   (D) 5

3. Which of the following statements is not true of RNA?

   (A) RNA is always found in the cytoplasm.
   (B) RNA is single stranded.
   (C) RNA contains the nitrogen base, uracil.
   (D) RNA contains the sugar, ribose.

4. In what part of a nucleotide is hereditary information carried?

   (A) sugar molecule
   (B) codon
   (C) phosphate group
   (D) nitrogen bases

5. The subunits that make up nucleic acids are _____?

    (A) amino acids
    (B) nucleotides
    (C) sugars
    (D) proteins

6. The form of nucleic acid that takes hereditary information from the nucleus to be copied in the cytoplasm is _____.

    (A) mRNA
    (B) rRNA
    (C) tRNA
    (D) DNA

7. You are working in a genetics lab and have been give a protein with the following sequence of nitrogen bases: UCU AUA GCC GGA. Using the mRNA codon chart on page 94, determine the sequence of amino acids.

    (A) lysine, phenylalanine, glycine, proline
    (B) aspartic acid, lysine, proline, alanine
    (C) methionine, proline, lysine, glycine
    (D) serine, isoleucine, alanine, glycine

8. Which of the following is not important to the process of genetic transcription?

    (A) accurate matching of codons to anticodons
    (B) accurate assembly of mRNA
    (C) careful selection of gene to be copied
    (D) all of the above are important in transcription

9. Which of the following mRNA codons would cause protein synthesis to end? (Refer to the mRNA codon chart on page 94)

    (A) CCC
    (B) UGA
    (C) GCA
    (D) UGU

10. Assembly of a protein is performed, repeatedly attaching amino acid to amino acid, during the process of _____.

    (A) replication
    (B) transcription
    (C) conservation
    (D) translation

Free-Response

1. Explain how nucleic acids are a universal physical feature shared by all living things.

2. How do nucleic acids provide us with evidence for evolution?

3. How do DNA and RNA direct the formation and assembly of a living organism?

Answer Explanations

Multiple-Choice

1. **C** The DNA strand TTA TAT GCA TGG CGC is complementary to the original segment of DNA which is **AAT ATA CGT ACC GCG.**

2. **D** In the DNA strand seen in Question #1, five amino acids will be added as a result of this sequence. Each of the codons represents one amino acid.

3. **A** The statements, "RNA is always found in the cytoplasm," is not true of RNA.

4. **D** Hereditary information is carried in the nitrogen bases.

5. **B** The subunits that make up nucleic acids are nucleotides.

6. **A** The form of nucleic acid that takes hereditary information from the nucleus to be copied in the cytoplasm is mRNA.

7. **D** A protein with the following sequence of nitrogen bases: UCU AUA GCC GGA has the amino acid sequence: serine, isoleucine, alanine, glycine.

8. **A** Accurate matching of codons to anticodons is not important to the process of genetic transcription; it is very important during translation.

9. **B** The mRNA codon UGA would cause protein synthesis to end.

10. **D** Assembly of a protein is performed, repeatedly attaching amino acid to amino acid, during the process of translation.

Free-Response Suggestions

1. **Explain how nucleic acids are a universal physical feature shared by all living things.**

   Nucleic acids are present, in the form of DNA and/or RNA in all life forms. In prokaryotes (bacteria), the nucleic acids are not contained inside a nucleus and are relatively simple but they are there. In all eukaryotes (Protista, Fungi, Plants, and Animals) nucleic acids can also be found, usually inside a nucleus. In all of these living things, nucleic acids provide the information for reproduction of both the organism and its cells.

   All living things share specific characteristics. These include growth, development, organization, homeostasis, metabolism, cellular construction, and reproduction. Reproduction in all living things originates in DNA. DNA provides the instructions to assemble amino acids in the correct order to accurately form needed proteins and to direct those proteins to do their jobs.

2. **How do nucleic acids provide us with evidence for evolution?**

   Natural selection and evolution are processes that are linked to DNA because DNA is the packaging that carries a parent's genetic information on to its offspring. When natural selection occurs, it's because some genetic element of a parent worked to its advantage and allowed the parent to mature and then to produce its own offspring. Those offspring that were then endowed with the parent's selective advantage are more likely to mature, reproduce, and pass that same trait on to their own offspring. After many generations, this trait with its selective advantage, is likely to be established as a characteristic of the population.

3. **How do DNA and RNA direct the formation and assembly of a living organism?**

   The order of the nitrogen bases, in the DNA of every organism, directs the assembly of the parts of that organism and the jobs to be performed by each part. The sequence of A, T, C, and G in DNA determines the sequence of A, U, C, and G in transcription of RNA. The mRNA sequence of A, U, C, and G forms codons that call for particular amino acids to be assembled into a specific order to make a specific protein. Ultimately, in translation, the order of nitrogen bases generates the correct assembly of proteins. Since living organisms are made primarily of protein, DNA is the guide for the correct assembly of all living things.

CHAPTER 7

Mendelian Genetics

Outline for Organizing Your Notes

I. Mendelian Inheritance
   A. Gregor Mendel
      1. History
      2. Work with Sweet Pea Plants
   B. Simple Mendelian Inheritance
      1. Alleles
      2. Dominant and Recessive
      3. Genotypes and Phenotypes
      4. Homozygous and Heterozygous
      5. Monohybrid and Dihybrid
   C. Punnett Squares and Pedigrees

II. Complex Patterns of Inheritance
   A. Incomplete Dominance
   B. Codominance
   C. Multiple Alleles
   D. Polygenic Inheritance
   E. Epistasis

III. Genetic Disorders
   A. Recessive Disorders
      1. Cystic Fibrosis
      2. Sickle Cell Disease
      3. Tay-Sachs

   B. Dominant Disorders
      1. Huntington's Disease
      2. Marfan Syndrome
   C. Sex Linked Disorders
      1. Hemophilia
      2. Duchenne Muscular Dystrophy
   D. Polygenic Disorders
      1. Cancer
      2. Autoimmune Disease
      3. Diabetes
   E. Chromosomal Abnormalities
      1. Trisomy 21
      2. Sex Chromosome Abnormalities

IV. Dealing with Genetic Disorders
   A. Ultrasound
   B. Amniocentesis
      1. Karyotyping
   C. Chorionic Villus Sampling
   D. Genetic Counseling

How can anyone make an accurate prediction about the traits of a baby that hasn't been born yet? Why do some members of families look so much alike while others look so different? How can two healthy parents have both a completely healthy baby and also a baby with a terrible genetic disease?

Dozens of years before anyone understood the molecular nature of genetics, or even that genes existed, a gardening monk in Austria meticulously recorded the characteristics of offspring of sweet pea plants. This man, Gregor Mendel, ultimately explained the process of predicting the outward appearances of generations of pea plant offspring. His work resulted in our earliest understanding of how traits are passed from parents to their young.

Gregor Mendel

Gregor Mendel (1822–1884) is considered to be the "Father of Modern Genetics." Mendel, an Austrian monk, worked as a teacher of physics, but when he wasn't teaching, he was gardening. From 1856 until 1863, Mendel paid special attention to the variations in sweet pea plants in his garden. He kept track of the following seven visible and distinct traits:

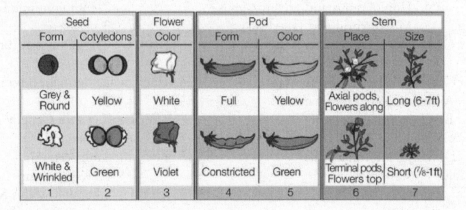

Seed		Flower	Pod		Stem	
Form	Cotyledons	Color	Form	Color	Place	Size
Grey & Round	Yellow	White	Full	Yellow	Axial pods, Flowers along	Long (6-7ft)
White & Wrinkled	Green	Violet	Constricted	Green	Terminal pods, Flowers top	Short (⅞-1ft)
1	2	3	4	5	6	7

Mendel controlled pollination by personally rubbing the pollen from one plant on the pistil of another plant and carefully documented the traits of many generations of his pea plants. He reported his findings in 1865 and published them in 1866. Without knowing about DNA or genes, because this information hadn't yet been discovered, Mendel described the way that characteristics are passed on from parents to their offspring and was able to explain the probability of a trait's appearance.

Simple Mendelian Inheritance

The first thing to remember about how traits are inherited is that our autosomal (body) cells contain two matched copies of every chromosome (46 chromosomes in 23 pairs). Each chromosome pair contains matched genes, and each of these paired genes is an **allele** to the other member of the pair. We receive the information in our chromosomes from the egg and sperm cells that united to make us. Following meiosis, gametogenesis, and fertilization, a healthy human **embryo** contains 23 pairs of chromosomes. We have a total of 46 chromosomes; 44 of them (22 pairs) are **autosomes** and one pair are the sex chromosomes, X and Y. Each pair of chromosomes is homologous, meaning that both members of the pair contain all the same genes but not necessarily the same information for those genes. So, for example, one parent might contribute the gene for polydactyly (extra fingers and toes) while the other parent's homologous gene may be for five fingers and toes. Only one of these two conditions will happen in each of their offspring, and which it will be is genetically predictable. It has been established that polydactyly, in the absence of other congenital abnormalities, is a **dominant** trait and that if this gene is passed from parent to child, the child will be born with extra digits. Dimples, albinism, male baldness, freckles, and nearsightedness are all single gene, dominant traits. The presence of five fingers on each hand and five toes on each foot is a **recessive** trait. Blue eyes, type O blood, and smooth hair are all recessive traits, so they are only seen when the dominant gene is absent.

When we talk about Mendelian inheritance, we are considering several related generations. The original generation is labeled "P" for parent generation. Their children are called the first filial generation, so the offspring of the parent generation is called "F_1". The next generation, the grandchildren of the P generation, are called the second filial, or "F_2" generation. Dominant traits are likely to appear in each generation. Recessive traits may seem to come out of nowhere, and in fact, are usually hidden in an F_1 generation.

A dominant trait is represented by a capital letter, a recessive trait is represented by the same letter in a lower case, and the capital letter always gets printed first. We could represent the dominant gene for polydactyly with a D and the recessive gene for five digits with a d. Each of the two homologous alleles in each individual will contain either a gene for D (polydactyly) or d (five digits). The possible combinations of alleles are called **genotypes**, and would be represented by DD, Dd, or dd. When both alleles are the same, they are **homozygous** and when they are different, they are **heterozygous**; so, the genotype DD is homozygous dominant, Dd is heterozygous, and dd is homozygous recessive. The condition, or appearance, that results from a genotype is a **phenotype**. The homozygous dominant and heterozygous offspring will both have the polydactyl phenotype and the homozygous recessive offspring will have a phenotype with five digits on each hand and foot. A heterozygous parent (Dd) has the genes for both the dominant and the recessive trait; so, as gametes are formed, each of this parent's gametes has a 50 percent chance of receiving the dominant allele and 50 percent chance of getting the recessive one.

As a result of his understanding of single trait (**monohybrid**) inheritance, and although meiosis and gametogenesis were not yet understood, Mendel was able to develop the law of segregation. This scientific law states that during gamete formation, alleles are separated from each other and then they are randomly reunited at the time of fertilization. As a result of his work with the inheritance of multiple traits (two traits are called a **dihybrid** cross and three traits are a trihybrid cross), Mendel developed the Law of Independent Assortment. This law states that each trait is passed to each offspring independently of any other traits.

Punnett Squares and Pedigrees

There are diagrams available to help us track the inheritance of traits and to make predictions for their occurrence. Let's consider the pattern of inheritance for a single trait. We can complete a Punnett Square to predict both genotype and phenotype. Generally, we write the male (m) genotype across the top of the square and we write the female (f) genotype down the side. We read the male alleles from the top of the Punnett Square to the bottom and we read the female alleles from left to right. In the following example, both parents have the heterozygous genotype (Gg) and the color yellow (G) is the dominant phenotype while the color green (g) is recessive. We assign one allele from each parent to each box in the Punnett Square. Remember, capital letters always come before lowercase letters.

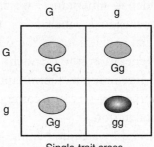

Single-trait cross
G = yellow, g = green

In this monohybrid cross there is a 25 percent chance that each offspring of these offspring will have a homozygous dominant (GG) genotype, 50 percent chance that each will be heterozygous (Gg), and a 25 percent chance that each will be homozygous recessive (gg). These genotype percentages tell us that there is a genotypic ratio of 1:2:1 (1GG:2Gg:1gg). There is a 75 percent chance that each offspring will have a yellow phenotype and a 25 percent chance that each offspring will be green. This gives us a phenotypic ratio of 3:1 (3 yellow: 1 green).

When we develop Punnett Squares for multiple traits they can get pretty big. A Punnett Square for a dihybrid cross will have 16 spaces and a trihybrid cross will have a Punnett Square with 64 spaces. Using these diagrams we can make predictions about genotypes and phenotypes. Always keep in mind that each box within a Punnett Square will only contain two copies of each homologous allele, one from mom and one from dad. When we fill in the boxes we take the one copy of each letter that is written at the top above that space and the one copy of each letter that is written to the far left of that space.

♂ **Gametes**

	RY 1/4	Ry 1/4	rY 1/4	ry 1/4
RY	RRYY 1/16	RRYy 1/16	RrYY 1/16	RrYy 1/16
Ry	RRYy 1/16	RRyy 1/16	RrYy 1/16	Rryy 1/16
rY	RrYY 1/16	RrYy 1/16	rrYY 1/16	rrYy 1/16
ry	RrYy 1/16	Rryy 1/16	rrYy 1/16	rryy 1/16

♀ **Gametes**

Legend:
- Round yellow
- Round green
- Wrinkled yellow
- Wrinkled green

In the dihybrid cross above, both parents are heterozygous (RrYy) for both traits. Because they are heterozygous for both traits the dominant trait will be expressed so they will both have the round and yellow phenotype. When these parents produce offspring together, it is represented by the phrase RrYy × RrYy. So, how do we fill out this Punnett Square? Well, first let's remember a few rules: When we're filling in the spaces to represent offspring, we always write both copies of the same letter together (for example, rr and YY), capital letters (representing dominant traits) come before lowercase (recessive traits), and we write the letters in alphabetical order (so it's written rrYY). When two parents who are heterozygous for two traits produce offspring, the phenotypic ratio of their dihybrid cross is 9:3:3:1. This means that there is a 9/16 chance that the offspring will exhibit both dominant traits, 3/16 chance that they will exhibit a dominant first trait and a recessive second trait, 3/16 chance that they will exhibit a recessive first trait and a dominant second trait, and a 1/16 chance that they will exhibit both recessive traits.

So, now do you want to see how a trihybrid cross would look? Here you go:

	ABC	**ABc**	**AbC**	**aBC**	**Abc**	**aBc**	**abC**	**abc**
ABC	AABBCC	AABBCc	AABbCC	AaBBCC	AABbCc	AaBBCc	AaBbCC	AaBbCc
ABc	AABBCc	AABBcc	AABbCc	AaBBCc	AABbcc	AaBBcc	AaBbCc	AaBbcc
AbC	AABbCC	AABbCc	AAbbCC	AaBbCC	AAbbCc	AaBbCc	AabbCC	AabbCc
aBC	AaBBCC	AaBBCc	AaBbCC	aaBBCC	AaBbCc	aaBBCc	aaBbCC	aaBbCc
Abc	AABbCc	AABbcc	AAbbCc	AaBbCc	AAbbcc	AaBbcc	AabbCc	Aabbcc
aBc	AaBBCc	AaBBcc	AaBbCc	aaBBCc	AaBbcc	aaBBcc	aaBbCc	aaBbcc
abC	AaBbCC	AaBbCc	AabbCC	aaBbCC	AabbCc	aaBbCc	aabbCC	aabbCc
abc	AaBbCc	AaBbcc	AabbCc	aaBbCc	Aabbcc	aaBbcc	aabbCc	aabbcc

Can you see that this is a cross between two parents who are each heterozygous for three traits? Can you figure out what the parent genotypes were? This cross represents the offspring of AaBbCc × AaBbCc. It should follow all the same rules as a dihybrid cross: Both A and a come before B and b, which come before C and c, and capitals are written before lowercase letters.

Another way to look at inheritance patterns is with a pedigree. Here is a pedigree that allows us to trace a family history as well as a disorder seen in some members of this family.

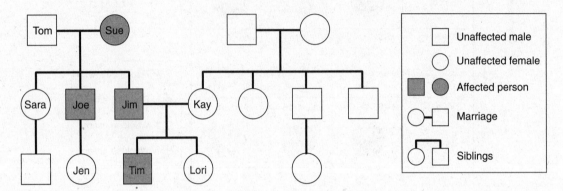

You can tell a lot from the information in this pedigree. You can tell that this is the family history of a married couple, Jim and Kay. The pedigree includes Jim's and Kay's parents, all of Jim's and Kay's siblings, and all of their children. You can identify the males (squares) and the females (circles) and you can see who is affected by an illness. You can trace a disorder through Jim's side of the family and can expect that if this is a genetic disorder; it is probably a dominant disorder (because it's seen in each generation and only one parent needs to have the gene) and it's not likely to be sex-linked (you know this because it's seen in both males and females).

Complex Patterns of Inheritance

Mendel was able to explain simple inheritance patterns based on the sweet pea plant, but there are also other, more complex, inheritance patterns to understand. **Incomplete dominance** is an inheritance pattern in which the heterozygous phenotype is an intermediate of the homozygous dominant and homozygous recessive. An example of incomplete dominance is seen in the colors of a four o'clock plant. The flowers of a homozygous dominant four o'clock plant are red and those of the homozygous recessive are white. In this plant the heterozygous phenotype is a pink flower. If a homozygous dominant and a homozygous recessive flower reproduce together, the F_1 generation will all be pink and the F_2 generation will have a 1red: 2pink: 1white ratio.

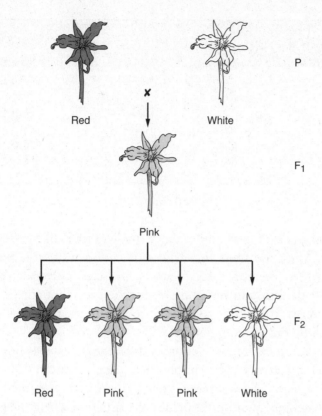

Codominance is a shared inheritance pattern, when heterozygous alleles are each expressed completely, not in an intermediate phenotype. Roan horses and cattle, as well as mixed-color patterns in snakes, mice, and alpaca, are all examples of two different homologous alleles being fully expressed. Codominance of color genes often appears in patterns of each different color hair, skin, or each allele's color in multicolored flowers. The visible, or phenotypic, difference between incomplete dominance and codominance is that with incomplete dominance the organism has an intermediate blend of the dominant traits (red + white = pink) and with codominance the organism has both dominant traits fully expressed (red + white = some red + some white).

Multiple alleles are a form of codominance. When there are more than two possible versions of an allele, offspring will only get two of the possible variations and generally they will each be expressed completely. Blood types, with three possible alleles, are a type of multiple allele codominance. Blood types i^A and i^B are both dominant alleles and type i is recessive so there are more possible genotypes and phenotypes.

Blood Type	Genotype		Can Receive Blood From
A	$i^A i$	AA	A or O
	$i^A i^A$	AO	
B	$i^B i$	BB	B or O
	$i^B i^B$	BO	
AB	$i^A i^B$	AB	A, B, AB, O
O	ii	OO	O

Blood Type Chart

Polygenic, or multiple gene, inheritance involves traits like eye color, hair color, and skin color. These are traits that have tremendous variation and can be influenced by two, three, or more genes. Cancer, diabetes, cleft palate, schizophrenia, congenital heart defects, and many more disorders are thought to have polygenic inheritance patterns. It is often possible to recognize clear degrees of difference in traits controlled by polygenic inheritance.

Epistasis is a result of the interaction of different genes. It is not the interaction of different alleles, that would be simple Mendelian inheritance, the interaction of dominant and recessive genes. Epistasis refers to the action of one gene to override another gene. Examples of epistasis include the situation where the gene for balding would mask the gene for a widow's peak, or the gene for albinism would override the gene for hair color.

Genetic Disorders

Inheritance patterns are interesting when we think of how we ended up looking the way we do. However, our understanding of genotypes and phenotypes can also influence life and death situations and decisions. Consider that while our genes direct our blood type, they also direct the formation of our blood cells. A gene with instructions for misshapen red blood cells is the cause of sickle cell disease. A gene directing the incorrect formation of clotting factor in the blood causes hemophilia. A gene that directs the incorrect function of sodium and chloride channels in cell membranes causes cystic fibrosis.

More than 6,000 different single gene genetic disorders have been identified and 1 in every 200 babies born in the United States has some type of genetic disorder. Single trait genetic disorders are the phenotypes that result from just two alleles, one from a mother and one from a father. They can be either dominant or recessive and can be passed down from the parents or can be the result of spontaneous mutations within the parents' gametes. There are also multifactoral disorders which can be caused by a combination of environmental factors, infectious processes, aging, random or unknown reasons, and genetic misinformation.

A recessive disease or disorder may be passed on when both parents are carriers of a disease gene. A parent that has one healthy allele and one disease-causing allele is called a carrier and has a 50/50 chance of passing that allele to his, or her, child. If two carriers produce an offspring, each parent will either contribute one allele that carries the disease or one healthy allele. The genotypic (and phenotypic) possibilities are 25 percent chance of DD (two healthy alleles and a healthy child), 50 percent likelihood of Dd (one healthy allele and one disease-producing allele and the child will be a carrier and may or may not be affected by the disease), or 25 percent dd (two disease alleles and the child will have the disease).

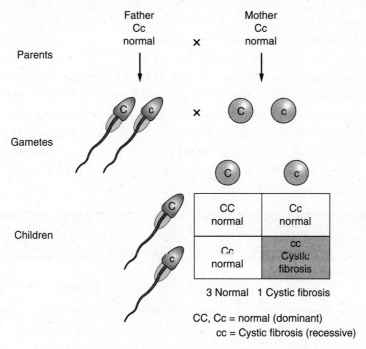

Typical Inheritance of a Recessive Disorder: Cystic Fibrosis

Let's consider several recessive genetic disorders caused by a single faulty gene:

- **Cystic Fibrosis** is one of the most common genetic disorders in the United States. CF causes abnormal ion transport across cell membranes that results in the development of thick, sticky mucous in the respiratory, digestive, and reproductive systems of its victims. Babies with cystic fibrosis have repeated respiratory problems, salty skin, and they tend to eat well but have difficulty gaining weight.

- **Sickle Cell** is another relatively common genetic disorder, most common in people of African descent. A homozygous recessive sickle cell patient will experience anemia (a low red blood cell count resulting in a reduction in oxygen availability to cells) and pain caused by red blood cells that are crescent-shaped; they may clump together causing clots to form and do not carry sufficient oxygen. One interesting feature of sickle cell inheritance is **heterozygous superiority**, in which a person who is heterozygous for sickle cell will have minimal sickle cell effects and will also have a resistance to malaria. Since malaria can be a life threatening infection and is common in equatorial regions, it is to an individual's advantage to carry the gene for sickle cell.

- **Tay-Sachs** is a rare genetic disorder that causes incorrect manufacture of an enzyme called Hexosaminidase-A (Hex-A). A baby with Hex-A deficiency will seem fine for its first six months of life but as lipids begin to accumulate in its brain, the infant begins to deteriorate. This child will go blind and deaf and will usually die by the age of five. One common diagnostic sign of Tay-Sachs is a red spot on the child's retina, on the inside and back of the eye. Tay-Sachs is found mostly in people of Ashkenazi Jewish descent and is incurable.

A dominant genetic disorder will be caused by a single disease-causing gene out of the four possible alleles available from the two parents. In other words, anyone who receives a single gene for a dominant disorder will have that disorder. These genes may be passed from an affected parent to their offspring or as a result of a mutation in an allele of a gamete. The inheritance pattern of a dominant disorder is typically 50 percent Dd (heterozygous, child has the disorder) or 50 percent dd (homozygous recessive, child unaffected).

		Father's Gametes	
		D	d
Mother's Gametes	d	Dd	dd
	d	Dd	dd

Typical Inheritance of a Dominant Disorder: Huntington's Disease

Some genetic disorders caused by a dominant gene include the following:

- **Huntington's disease** causes mental deterioration, personality changes, and severe body tremors. This disorder is generally late onset with mild symptoms beginning around the age of 40. The patient's condition will deteriorate with death occurring from 10 to 30 years after the onset of symptoms. There is no cure for this disease. One of the unique features of this disorder is that its symptoms usually don't appear until people have had their children and potentially passed the gene on to the next generation. With the advent of genetic testing, the children of Huntington's victims now have the option of finding out whether this degenerative condition is part of their own destiny.

- **The Marfan syndrome** is a disorder of the body's connective tissue. It causes the victim to manufacture a weakened version of a connective tissue called fibrillin. Marfan's patients tend to be disproportionately tall and lean; in fact, their arm span length is usually longer than their height. They often have nearsightedness and may have to avoid contact sports and roller coasters to prevent lens dislocation. They can often hyperextend their joints (they are often double-jointed), may have flat feet, and a protruding or indented sternum (breastbone). The biggest risk for Marfan's patients is aortic dissection, or splitting of the aorta, which can be fatal.

Once diagnosed, a Marfan's patient can be monitored and treated before emergencies occur and as a result, can have a great prognosis for a normal life span.

Sex-linked genetic disorders are carried on sex chromosomes. Remember, a female has two X chromosomes (XX) and a male has an X and a Y (XY). The X chromosome is larger than the Y chromosome and tends to be the one that carries most sex-linked disorders. In men, a recessive disorder trait on the top of an X chromosome doesn't have a second, healthy recessive gene to prevent its expression. We write symbols for sex chromosome alleles differently because the X and Y influence the outcome. For example, the recessive trait for hemophilia is carried on the X chromosome; X^H represents the X chromosome with the healthy blood allele and X^h represents the X with the hemophilia allele.

		Genotype (phenotype)					Genotype (phenotype)		
♀		X^HY (normal)		♂		♀	X^hY (hemophiliac)		♂
Gametes		X^H	Y	Gametes	Gametes		X^h	Y	Gametes
X^HX^h (carrier)	X^H	X^HX^H (normal)	X^HY (normal)		X^HX^h (carrier)	X^H	X^HX^h (carrier)	X^HY (normal)	
	X^h	X^hX^H (carrier)	X^hY (hemophiliac)			X^h	X^hX^h (hemophiliac)	X^hY (hemophiliac)	

Punnett squares for the sex-linked trait hemophilia

The following are sex linked recessive disorders:

- **Hemophilia** is a blood clotting disorder mostly seen in males; someone with any form of hemophilia who gets a cut will take longer to clot than someone with normal clotting factors. Hemophiliacs often bleed into their joints, their muscles, or from their digestive tracts; this may cause pain, weakness, swelling, and anemia. Treatment for hemophilia usually involves the infusion of plasma derived from human blood or the infusion of a genetically engineered clotting factor.

- **Duchenne Muscular Dystrophy** is another sex-linked disorder and therefore, affects mostly boys. The gene that is not working correctly codes for a structural component of cell membranes in muscle tissue. In DMD, this causes muscular degeneration beginning between infancy and six years of age. Muscular deterioration begins in the legs and pelvis and most boys with DMD are wheelchair bound by age 12; skeletal deformities and paralysis usually follow. While recent medical advances have increased the life expectancy and improved the quality of life for a DMD patient, ultimately, a muscular degeneration of muscles involved with the heart and with breathing has kept the life expectancy of DMD patients at late teens to mid-30s (with some patients surviving into their 50s).

Multifactoral or polygenic disorders are disorders that occur because of interactions between multiple genes, the environment, and an individual's lifestyle. These disorders are not inherited through simple Mendelian patterns but can often be traced through families, lifestyle choices, and environmental exposures.

- There are many types of cancer but the bottom line is that cancerous cells have lost the controls that keep them dividing only when new cells are needed and the controls that tell them to die off when they get too old. This cellular immortality causes cancer, or malignancy, to develop. Research has indicated that interactions between genes, the environment, and lifestyle can trigger cellular immortality. For example, overexposure to radiation, to certain chemicals, to certain diets, triggers the development of cancer in some people but not others.

- Autoimmune Diseases including multiple sclerosis, lupus, and rheumatoid arthritis, all seem to have a hereditary component. It has been suggested that a number of autoimmune disorders are linked to a large gene complex on chromosome six. While the mechanisms are not completely understood, many autoimmune diseases can be tracked through family histories.

- The development of both Type I and Type II diabetes appear to require a genetic predisposition and then some environmental trigger to initiate the disease process. Environmental triggers for Type I diabetes include possible viral infections, cold weather, breastfeeding, and for Type II diabetes, there is a clear link between genetics and diet.

Chromosomal abnormalities are different from genetic defects; chromosomal abnormalities refer to the entire chromosome rather than a little piece of it.

- Trisomy 21 is another name for Down syndrome. It is the result of an incorrect division of chromosomes during gamete formation, with one gamete getting an extra chromosome 21. A baby that develops from this gamete will end up having 47 chromosomes instead of 46.

  Symptoms of Down syndrome include varying degrees of intellectual disabilities, wide set eyes, a thick tongue, and, often, heart defects. There are other types of trisomy; the birth defects of Trisomies 13 and 18 are much more severe than Trisomy 21 and are often fatal in the first year of a baby's life.

- Klinefelters (XXY), XYY syndrome, Triple X (XXX), and Turner (XO) syndromes are all chromosomal abnormalities of the sex chromosomes. Some phenotypes (XXX and XYY syndromes) are relatively indistinguishable from the normal XX or XY phenotypes. Klinefelters and Turner syndromes may have abnormalities of the reproductive system with a likelihood of fertility problems.

Predicting and Modifying Outcomes

Women who are expecting babies may choose to have tests performed to let them know how their baby is developing and if there are likely to be any serious problems with the baby they're carrying. This may be because a woman is older than a certain age and therefore, has a higher risk of conceiving a baby with chromosomal abnormalities, if there is a family history of genetic or birth defects, or if routine prenatal screenings of blood and urine indicate a potential problem.

Ultrasound is a test performed during many pregnancies in America today. Ultrasound uses sonar technology to produce an image of internal structures. It is a non-invasive (the skin isn't cut or penetrated) way to establish the developmental progress of a baby. An ultrasound can also tell gestational age (the baby's due date), if there are multiple babies, if there are fetal deformities, and the location of the placenta. Other, more invasive tests, often use ultrasound imaging to guide instruments away from harming the fetus.

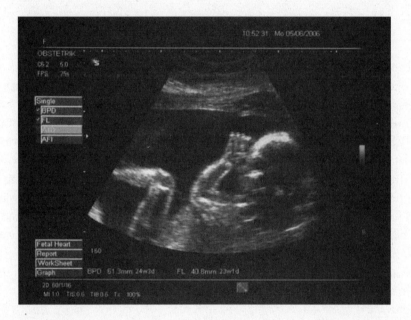

Amniocentesis is a test that begins with the withdrawal of about two tablespoons of amniotic fluid from a pregnant woman's uterus usually around 16 to 20 weeks into her pregnancy. Amniotic fluid is the liquid that surrounds the developing baby. This test must wait until there is enough amniotic fluid surrounding and protecting the fetus. This doesn't happen until approximately the 16th week of a pregnancy. The test is done using a needle inserted through the mother's abdomen into the uterus, aimed with the help of an ultrasound image. Amniotic fluid contains cells that have been given off by the fetus that can be used to establish a genetic and chromosomal profile for the baby. The amniotic fluid is sent to a lab where it is cultured and grown for ten days. At this point a chromosome study, or **karyotype**, can be accomplished. A karyotype allows the chromosomes to be counted and identified; it also allows abnormalities of the chromosomes to be recognized.

Normal Human Male Karyotype

Another type of prenatal testing is called **chorionic villus sampling**, or CVS. This procedure is performed earlier in a pregnancy, usually around the 10th to 12th week. It involves taking a biopsy of placental cells called chorionic villi. CVS provides results of chromosomal and genetic testing weeks earlier than amniocentesis. Both CVS and amniocentesis create a small percentage chance for a spontaneous miscarriage due to a compromised "seal" of the sterile intrauterine environment.

When there is a family history, or genetic disorder, or when results come in from ultrasound, CVS, or amniocentesis that indicate a birth defect, parents are often referred for **genetic counseling**. A genetic counselor is trained and certified to help people who are impacted by a genetic disorder; they may act as an advocate for parents, patients, and families, they may be involved in identifying and educating families at risk, analyzing familial inheritance patterns, evaluating potential for recurrence within a family, and exploring available testing options.

Important Terms to Remember

Alleles: Alternative forms of a gene on matched, or homologous, chromosomes.

Amniocentesis: A prenatal test to evaluate the likelihood of genetic or chromosomal disorders in a developing fetus. Test is performed by withdrawing a small amount of amniotic fluid through the mother's uterus and then culturing the fluid to evaluate fetal cells.

Autosome: Any chromosome that is not a sex chromosome; humans have 22 pairs of autosomes.

Chorionic villus sampling: A biopsy of placental tissue; allows early diagnosis of genetic and chromosomal abnormalities.

Codominance: A shared inheritance pattern; when heterozygous alleles are each expressed completely, not in an intermediate phenotype.

Dihybrid: Two gene inheritance pattern; represented by a double genotype (DdMM or DDmm).

Dominant: A version of a gene, or allele, which is expressed in offspring. In Mendelian inheritance, a dominant gene prevents the expression of a recessive gene.

Embryo: A fertilized egg that has begun dividing; in humans, embryonic development lasts from fertilization through week eight of a pregnancy.

Epistasis: The action of one gene to override another gene.

Genetic counseling: A profession that assists all parties impacted by genetic disorders by helping families understand genetic disease transmission, advocating for families affected by genetic disorders, and helping them to understand risk for genetic disease transmission.

Genotype: A combination of alleles; these alleles may be dominant or recessive.

Heterozygous: A combination of alleles; one allele will be dominant and one will be recessive. In simple Mendelian inheritance patterns, the phenotype will be dominant.

Heterozygous superiority: A genetic scenario in which it is preferable to be heterozygous for a genetic disorder. Inheritance of sickle cell has this peculiarity, a person who is heterozygous for sickle cell will have minimal sickle cell effects and will also have a resistance to malaria.

Homozygous: When both alleles are the same; either both are dominant or both are recessive.

Incomplete dominance: An inheritance pattern in which the heterozygous phenotype is an intermediate of the homozygous dominant and homozygous recessive.

Karyotype: A chromosome study; often performed following amniocentesis to evaluate the genetic and chromosomal integrity of a developing fetus.

Monohybrid: Single gene inheritance pattern; represented by a single genotype (MM, Mm, or mm).

Multiple alleles: A form of codominance. When there are more than two possible versions of an allele, offspring will only get two of the possible variations and generally they will each be expressed completely. An example is blood typing.

Phenotype: The expression of a genotype; phenotype is the way that an organism shows its genotype.

Polygenic: Multiple gene inheritance that involves complex traits like hair color and skin color. These traits that have tremendous variation and can be influenced by two, three, or more genes.

Recessive: The version of a gene, or allele, which will not be expressed in the presence of a dominant gene.

Ultrasound: A technology that allows imaging of internal body structures; it is used frequently to evaluate fetal progress and development.

Practice Questions

Multiple-Choice

1. You are working in a plant nursery, repotting snapdragons for sale. You notice that 75 percent of the flowers are red and 25 percent are white. You are observing the plants' _____.

   (A) heterozygous superiority
   (B) chromosomes
   (C) genotype
   (D) phenotype

2. When Mendel was observing inheritance patterns in sweet pea plants and he mixed a homozygous dominant with a homozygous recessive P generation, what genotypes occurred in the F_1 generation?

   (A) all homozygous dominant
   (B) all heterozygous
   (C) all homozygous recessive
   (D) 50 percent homozygous dominant and 50 percent homozygous recessive

3. When Mendel mixed any two members of the F_1 generation in Question #2, what was the genotype of the next generation?

   (A) 1:1:1:1
   (B) 2:2
   (C) 1:2:1
   (D) 3:1

4. Your aunt is pregnant. Her doctor says she should have a test called a(n) _____ because she is over 35 years old and therefore, the risk of _____ is increased.

    (A) amniocentesis; Trisomy 21
    (B) x-ray; progeria
    (C) ultrasound; sickle cell disease
    (D) CVS; hemophilia

5. Material taken from the prenatal test in Question #4 will be grown and cultured for about ten days. It will then be treated to show mitotic chromosomes and allow them to be photographed. This process is called _____.

    (A) karyotyping
    (B) chorionic villus sampling
    (C) gene monitoring
    (D) ultrasound imaging

6. If your aunt's testing turns out to be positive, it is likely that the baby will be born with _____.

    (A) hemophilia
    (B) Marfan syndrome
    (C) Down syndrome
    (D) progeria

7. Your classmate's Uncle Jim seemed really healthy until around his 41st birthday, then she says he started getting forgetful and he seemed to have a lot of tremors. Now, four years later, he gets confused easily and seems like his entire body is always jerking and twitching. His doctor began genetic testing for _____.

    (A) Huntington's disease
    (B) Cystic Fibrosis
    (C) Marfan syndrome
    (D) Duchenne Muscular Dystrophy

8. Uncle Jim (from question) 7 has four kids. If his genetic tests come back positive, what is the probability that each of his kids will have the same disease?

    (A) 0 percent
    (B) 25 percent
    (C) 50 percent
    (D) 100 percent

9. Your best friend's older brother and his wife had a baby about five months ago. This baby seems to get a lot of colds and now she's in the hospital with pneumonia. When your friend was visiting, she overheard the doctor ask if the baby's skin was salty and the mom sounded surprised but said "yes." What is the doctor likely to test the baby for?

   (A) sickle cell disease
   (B) cystic fibrosis
   (C) Huntington's disease
   (D) Marfan syndrome

10. If a man has genes for black hair and a gene for male pattern baldness, he will become bald. This is an example of _____.

   (A) heterozygous superiority
   (B) codominance
   (C) multiple alleles
   (D) epistasis

Free-Response

1. Two of your cousins, who are sisters, have been diagnosed with an autosomal dominant disorder called nail patella syndrome. They both have underdeveloped thumbnails and kneecaps, poor kidney function, and scoliosis. How was this inherited? Draw and explain a pedigree of your extended family including these two relatives.

2. Gregor Mendel meticulously recorded the phenotypes of many generations of the sweet pea plant. He watched seven traits very carefully in this plant. Do you think his selection of the sweet pea plant and of these seven traits was a good, or bad, idea? Why?

3. You had to do a report on multiple sclerosis for your science class. Your research led you to several journal articles that said there was a relationship between a number of autoimmune disorders including MS, lupus, rheumatoid arthritis, and Type I diabetes. Discuss your explanation for this observation.

Answer Explanations

Multiple-Choice

1. **D** Whenever you're observing an organism's appearance, you're looking at its phenotype. There is no advantage in snapdragon color in this scenario, so no reflection of heterozygous superiority. You cannot see chromosomes with the naked eye and you can't see a genotype, you can only see genotype reflected in the phenotype.

2. **B** If a parent generation is HH × hh and you do a Punnett Square of the offspring (the F_1 generation) will be heterozygous (Hh).

3. **C** If you do a Punnett Square of the F_1 generation above, Hh × Hh you'll see a genotypic ratio of 1:2:1. This represents, and can be read, 1HH to 2Hh to 1hh.

4. **A** Amniocentesis is suggested to most pregnant women who are over 35 years old in the United States. Because of her age, her baby is at a higher risk of having Trisomy 21. X-rays are not recommended, for pregnant women and the risk of the baby having sickle cell or hemophilia is not increased because of her age.

5. **A** Karyotyping is the only process, of those listed, that is performed on cultured amniotic fluid, or that provides an opportunity to photograph mitotic fetal chromosomes.

6. **C** A positive test for Trisomy 21 indicates that the baby will have Down syndrome.

7. **A** This sounds like Huntington's disease; symptoms beginning after age 40, tremors, and forgetfulness. These symptoms do not match those of cystic fibrosis, Marfan syndrome, or DMD.

8. **C** Huntington's disease is an autosomal dominant genetic disorder and so for each of Uncle Jim's four children there is a 50 percent chance that they inherited the Huntington's disease allele from him.

9. **B** A baby with lots of colds and respiratory problems, and whose skin is salty, will probably be tested for cystic fibrosis. Sickle Cell, Huntington's disease, and Marfan syndrome all have different symptoms than this baby is experiencing.

10. **D** When the gene for baldness overrides the gene for hair color, it's called epistasis.

Free-Response Suggestions

1. **Two of your cousins, who are sisters, have been diagnosed with an autosomal dominant disorder called nail patella syndrome. They both have underdeveloped thumbnails and kneecaps, poor kidney function, and scoliosis. How was this inherited? Draw and explain a pedigree of your extended family including these two relatives.**

   Either your aunt or uncle must also have nail patella syndrome. One parent most likely passed the dominant allele on to both children. It's highly unlikely that this is a spontaneous mutation because two offspring both have the syndrome. So, the child of each pregnancy of your aunt and uncle will have a 50/50 chance of getting the gene and the disorder.

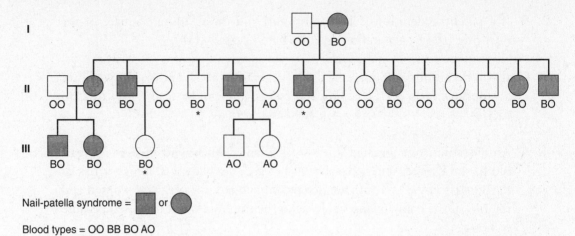

Nail-patella syndrome = ■ or ●

Blood types = OO BB BO AO

2. **Gregor Mendel meticulously recorded the phenotypes of many generations of the sweet pea plant. He watched seven traits very carefully in this plant. Do you think his selection of the sweet pea plant and of these seven traits was a good, or bad, idea? Why?**

Mendel used a type of plant that had rapid reproduction and very distinct and observable traits. Rapid reproduction is good for genetic experimentation because Mendel was able to control reproduction and inheritance patterns and he was able to record results for many successive generations. Distinct and observable traits are important because ambiguous results, or results that are difficult to assess, add an unnecessary variable to the experiment.

3. **You had to do a report on multiple sclerosis for your science class. Your research led you to several journal articles that said there was a relationship between a number of autoimmune disorders including MS, lupus, rheumatoid arthritis, and Type I diabetes. Discuss your explanation for this observation.**

Autoimmune diseases including MS, lupus, rheumatoid arthritis, and Type I diabetes have been linked to a complex of genes on chromosome 6. The right combination of genes along with environmental and lifestyle factors, can trigger any of these autoimmune disorders in different people, or may not cause any medical problem at all.

CHAPTER 8

Genetic Technology

Outline for Organizing Your Notes

I. Selective Breeding and Test Crosses

II. Manipulating DNA
    A. Restriction Enzymes
    B. Sticky Ends vs. Blunt Ends

III. Lab Work with DNA Fragments
    A. Gel Electrophoresis
    B. DNA Fingerprinting
    C. Polymerase Chain Reaction

IV. Gene Therapy

V. Genetic Engineering

VI. Exploring Genetics Further
    A. Genomics
    B. Proteomics

Have you ever wondered how we ended up with so many different types of dog, cat, horse, or rose? Have you ever considered the similarities and differences between a Balinese cat and a Bengal tiger? How about a Chihuahua and a wolf? How about all the different types of tulip or rose? Have you eaten any seedless watermelon, oranges, or grapes? Have you noticed that strawberries have become enormous and that you can get peaches without fuzz? Have you seen in the news that crimes are being solved, bodies are being identified, and diseases are being cured using DNA sequences? Can you imagine a day when a severely injured or diseased patient would be able to get replacement body parts from a genetics laboratory?

The human race has been manipulating genetics for many centuries, longer than they have understood genes or even known about heredity. Now we're looking at a future with almost limitless possibilities for human influence of genetic controls. Where will these alternatives take us and what are our responsibilities to ourselves and to future generations?

Selective Breeding and Test Crosses

From the time that humans began to populate the Earth, they've been altering their environment. By building shelter, tools, and fires, they've used available resources to make their world better for themselves. Another form of environmental manipulation is **selective breeding** of animals and plants. Early humans interacted with other living things in their habitats. As our ancestors recognized characteristics and behaviors of these other species that might be beneficial to the humans, the humans supported the survival of species and members of species that best fit their needs. This gave a selective advantage to organisms that best met human criteria for usefulness. As centuries went by, selected behaviors and characteristics became more and more refined. For example, a submissive wolf that stayed closed to a human ancestor's encampment 100,000 years ago, in an effort to have easy access to shelter and to the scraps left behind after a hunt, may have created a legacy of domestication that we see today.

There are more than 150 **breeds** of dog, 80 breeds of cat, 260 breeds of horse, 1,000 breeds of sheep, 800 breeds of cattle, 100 **cultivars** of tulip, and more than 6,500 varieties of rose. This tremendous diversity within species did not happen as a result of natural selection or simply because humans protected the species' ability to select for specific characteristics. For centuries, humans have controlled reproduction between members of species that had preferred traits, triggering improved milk production in cows, increasingly beautiful and fragrant flowers in roses, better wool production in sheep, larger and more flavorful fruit, and appropriate pet and protection characteristics and behaviors in dogs. When humans select organisms to be mated based on the characteristics that humans find favorable, a process known as selective breeding, many offspring will develop those same positive characteristics and then they are passed from generation to generation until they become typical for that species.

Until people gained an understanding of genetics they simply encouraged reproduction between individual members of a species that had preferred traits. With an understanding of simple Mendelian genetics, breeders and farmers can establish homozygous lines of descent thus strengthening those chosen characteristics. First, the breeder needs to know whether the selected trait is dominant or recessive. Once a trait is established as recessive, a breeder can conduct a **test cross**, breeding an organism with a dominant trait with an organism that exhibits the recessive version of that trait. The recessive organism is known to be homozygous. When offspring are produced, they will be 100 percent dominant if the dominant parent is homozygous for that trait. If the dominant parent is heterozygous, then half of the offspring will exhibit the dominant characteristic and half will exhibit the recessive.

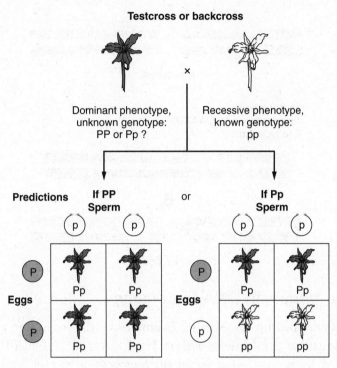

Testcross Possible Results

Manipulating DNA

Use and manipulation of genes is a huge part of science today. How does it work? First, remember that chromosomes are long strands of genes made up of DNA and that DNA is just nucleotide molecules (remember: sugar, phosphate group, and nitrogen base) attached to each other. If scientists want to investigate a gene, they will first need to isolate and identify the gene, or the sequence of nucleotides that code for a particular protein. This is done by cutting up the strand of DNA, using a type of biological scissors called a **restriction enzyme**. Restriction enzymes are made by bacteria to defend themselves against viral invaders; bacteria use these weapons to slice up and inactivate viruses. Scientists can take these restriction enzymes, identify the unique nitrogen base sequences that each will cut, and use them to slice through those same nitrogen base sequences in other types of organisms.

There are hundreds of restriction enzymes and each of them recognizes a single **restriction site**, a specific nitrogen base sequence that is four to eight base pairs long. Some restriction enzymes slice straight through the DNA at their restriction sites, producing a straight or blunt end. Other restriction enzymes cut a staggered path through the nitrogen bases, leaving an extension, called a sticky end, on one side of the DNA. Sticky ends tend to produce stronger connections when DNA strands are reconnected because there are more places for bonds to form.

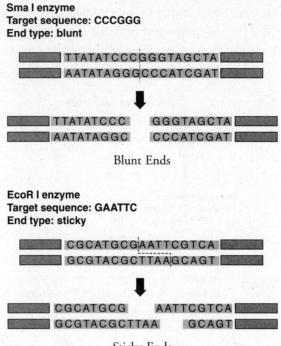

Blunt Ends

Sticky Ends

Laboratory Work with DNA Fragments

Once DNA is chopped up into smaller fragments by the restriction enzymes, there are several things that can be done with it. First, though, the DNA fragments need to be separated from each other based on fragment size. This separation can be done using **gel electrophoresis**, a process that begins with the application of small amounts of fragmented DNA into tiny wells in a layer of solidified gelatin. A negative electrode is positioned behind the wells and a positive electrode is introduced at the other end of the gel plate. Because the DNA is negatively charged, it will move across the gel plate toward the positive electrode. Different sized DNA fragments will move at varying speeds through the gel, their speed based on fragment size. Small fragments move through the gel faster than large fragments, there is much less resistance to their progress.

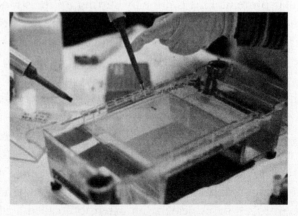

Gel Electrophoresis Plate

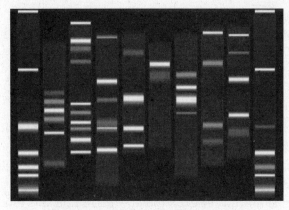

Gel Electrophoresis Restriction Map

Restriction maps are band patterns on the gel plate; the final visual result of running a gel electrophoresis. A restriction map shows us the DNA fragment lengths produced by cutting DNA with specific restriction enzymes. This process does not tell us what these sequences of bases do, but restriction maps can give us valuable information. Comparing unidentified restriction maps with known DNA sequences is a process known as **DNA fingerprinting**. DNA comparison techniques allow investigators to reunite children to their parents following abduction or natural disaster, to confirm or refute the identities of unidentified bodies and body parts, to better understand crime scenes, and to identify potential suspects in a criminal investigation. This process can be used to understand relationships between different species or to help classify a newly discovered organism. DNA comparison studies can also identify the presence of mutations and inherited conditions that cause genetic disorders.

One of the difficulties of relying on DNA evidence is that often there is very little DNA available to examine. This problem has been minimized with the advent of a technology called **polymerase chain reaction**, also known as PCR. PCR allows very small amounts of DNA to be copied millions of times in just a few hours. The process of PCR amplification involves several steps and only a few ingredients: the DNA (to be copied), lots of nucleotides (adenine, cytosine, guanine, and thymine), enzymes (DNA polymerases), and two **primers** (short strands of DNA that serve as starting and ending points for the new copies of DNA). First, the reactants are combined and heated to around 195°F for several seconds. This step separates the complementary strands of DNA. Second, the reactants are cooled to about 130°F, allowing the primers to bond with the complementary DNA strands. Third, the reactants are heated up to approximately 152°F, a temperature at which the DNA polymerases are most active and available to bond reactant nucleotides to their complementary nucleotides on the waiting DNA strands. Each time this cycle is repeated, the DNA content doubles. This series of steps is usually repeated 30 times to produce more than a billion copies of the original DNA fragment. Now there's plenty of DNA to examine!

Gene Therapy

When genes aren't working correctly as a result of a mutation or a genetic disorder, they are likely to produce the wrong protein or to make a nonfunctional protein.

This may result in disease and can be fatal. **Gene therapy** is an experimental procedure which manipulates genes in an effort to treat diseases without using surgery or medications. Currently in the United States, because it is considered a risky treatment option, gene therapy is only being used in response to disorders (certain single gene disorders, viral infections, and some cancers) for which there is no cure or alternative treatment. Most commonly, gene therapy replaces a defective version of a gene with a healthy and functional gene. Gene therapy can also involve the modification of gene regulation (increasing or decreasing a gene's activity), inactivating a mutated gene, or the introduction of a new gene to fight a disease or to produce a necessary protein.

There are two major kinds of gene therapy, germ line and somatic gene therapy. Germ line gene therapy inserts functional genes into sperm or egg cells that carry a genetic disorder. This type of gene therapy results in a change to the recipient's genes that can be passed on to future generations and will be present in all of the patient's body cells. Although this sounds like a great idea for eliminating many genetic disorders, there are still many technical, medical, ethical, and legal issues to be resolved before it becomes a commonly accepted practice. Somatic gene therapy is more commonly used and socially accepted because it targets only those body cells that are not working correctly; it doesn't become part of the patients germ cells or genome so it won't be passed on to their children. How does somatic gene therapy work? First, the target (dysfunctional) gene needs to be identified and its replacement, or therapeutic gene, isolated. The therapeutic gene is inserted into a carrier, or **vector**, which is often a virus that has been modified to contain normal human DNA. The vector is delivered to the target human cells and releases the healthy gene that it's carrying. If all goes well, the target cells incorporate the new DNA and begin working correctly. Somatic gene therapy has been used effectively in some cancers and in some inherited and acquired conditions of the kidneys, lungs, liver, and cardiovascular system.

Genetic Engineering

Genetic engineering is the term that describes direct manipulation of genes in a living organism. This process can involve the recombination of genetic material or the introduction of new DNA (from the same or a different species) into an organism. Genetically modified (GM) bacteria have been used to develop a variety of medications as well as antimicrobial products that target specific disease-causing bacteria like *E. coli* and *Salmonella*. GM bacteria are also being used to consume environmental toxins during oil and hazardous material spills, a process known as **bioremediation**. GM technology has been used in agriculture to produce disease-resistant and pest-resistant crops, to increase freshness, and to enhance the nutritional value of crops. Genetically modified soybean, cotton, corn, canola, sugar cane, sugar beets, and rice have been grown successfully while reducing the environmental impact of pesticides and fertilizers.

There are many ways that GM animals are being used today. Genetically engineered fish are thriving in farmed environments, increasing their value as a food source by growing large very quickly. Farmed animals in extreme environments may be genetically engineered to make them more resistant to heat, cold, drought, and disease. Milk cows are genetically modified with bovine growth hormone (BGH) causing them to produce a much larger amount of milk. When an organism is genetically modified by the insertion of DNA from another species into their DNA it is called a **transgenic organism**. A variety of animals are modified with human DNA to act as a source of cells, tissues, and organs for transplantation without the likelihood of rejection. In fact, the valves from the hearts of GM pigs and cows are used frequently in human heart valve replacement surgery. Many medications are produced in the bodies of GM animals and are called **biopharmaceuticals**. These drugs include hormones like insulin and estrogen, blood clotting factors, vaccines, growth factors and inhibitors, and interferon. Before these products became available through GM technology, they were often derived from the bodies of dead humans and animals. The process of extracting products to make pharmaceuticals was severely limited by the short supply of bodies and by the high cost of extraction and purification.

Exploring Genetics Further

Genomics is the study of the genomes of living things. A **genome** is the entire collection of genes in a full set of chromosomes for any organism. Interest in this field has existed since before 1972, when the first viral gene was sequenced. In 1995, the first full genome of a bacterium was sequenced. In the year 2000, the members of the Human Genome Project had sequenced the entire human genome. Scientists were surprised to learn that instead of human DNA consisting of the predicted 100,000 functional genes, *Homo sapiens* have only around 23,000 genes. A newer, and even more complex and current, field of research in science today is **proteomics**, the study of all the proteins expressed by all the genes in the genome.

Important Terms to Remember

Biopharmaceuticals: Medicines produced through biological technology.

Bioremediation: The use of biological compounds and living organisms, often genetically engineered, to degrade toxins in the environment and restore an ecosystem to its original, healthy condition.

Breeds: A further division of animal species; a subspecies; closely related organisms that can interbreed and produce fertile offspring, yet are very diverse and different within their species.

Cultivars: A further division of plant species; a subspecies; closely related organisms that can interbreed and produce fertile offspring, yet are very diverse and different within their species.

DNA fingerprinting: A process that involves comparing unidentified banding patterns with known DNA sequences to identify the source of the unknown DNA.

Gel electrophoresis: A mechanical process using gelatin and an electrical field to separate DNA fragments by size.

Gene therapy: An experimental procedure that manipulates genes in an effort to treat diseases without using surgery or medications, most often by replacing a defective gene with a healthy one.

Genetic engineering: The direct manipulation of genes in a living organism; also called gene splicing, recombinant DNA technology, and genetic modification.

Genome: The entire collection of genes in a full set of chromosomes for any organism.

Genomics: The study of the genomes of all living things.

Polymerase chain reaction (PCR): A process that allows very small amounts of DNA to be copied millions of times in just a few hours.

Primer: Short strands of DNA that serve as starting and ending points for new copies of DNA made during PCR.

Proteomics: The study of all the proteins expressed by all the genes in the genome.

Restriction enzyme: Catalytic protein made by bacteria that cuts DNA every time it recognizes specific nitrogen base sequences.

Restriction maps: After running a gel electrophoresis, the lines that show up on the gel indicate the lengths of the DNA fragments.

Restriction site: A specific nitrogen base sequence that is four to eight base pairs long that is recognized and cut by its unique restriction enzyme.

Selective breeding: Reproductive process involving selected organisms that are mated based on the characteristics that humans find favorable.

Test cross: Breeding a homozygous recessive organism with an organism that has a recessive trait. If the dominant organism is homozygous, all offspring will exhibit the dominant trait. If the dominant organism is heterozygous, half the offspring should exhibit dominant and half recessive traits.

Transgenic organism: An organism that is genetically modified by the insertion of DNA from another species into their DNA.

Vector: A structure that carries an infectious particle into its next host.

Practice Questions

Multiple-Choice

1. A breeder wants to figure out if a dog is homozygous for a dominant trait and will, therefore, pass on that trait to 100 percent of its offspring. The breeder can cross that dog with a _____ female dog. This test is called a _____.

   (A) heterozygous recessive; selection test
   (B) heterozygous dominant; test cross
   (C) homozygous dominant; selection test
   (D) homozygous recessive; test cross

2. In order to isolate a specific selected gene, it can be cut out by a protein called

   _____.

   (A) a viral knife
   (B) protein scissors
   (C) a restriction enzyme
   (D) a restriction site

3. A mechanism for separating pieces of DNA by fragment size is called _____.

   (A) gene manipulation
   (B) gel electrophoresis
   (C) PCR
   (D) DNA fragmentation

4. The driving force that pulls DNA fragments across a gel plate in question #3 is _____.

   (A) a positive electrode
   (B) a negative electrode
   (C) osmosis
   (D) endocytosis

5. Comparing the restriction map from a known source with one from an unknown source is known as _____.

   (A) comparative restriction
   (B) genetic engineering
   (C) DNA fingerprinting
   (D) PCR

6. A form of genetic engineering that develops bacteria that can consume and break down hazardous material and oil spills is called _____.

   (A) vector replacement technology
   (B) polymerase chain reaction
   (C) biogenic manipulation
   (D) bioremediation

7. Genetically modified drugs including hormones, clotting factors, and vaccines are all types of _____.

   (A) biopharmaceuticals
   (B) biomedication
   (C) biological mediation
   (D) bioremediation

8. When the Human Genome Project was completed, scientists were surprised to find that the human genome included approximately _____ genes.

   (A) 12,000
   (B) 23,000
   (C) 43,000
   (D) 100,000

9. The first gene ever sequenced belonged to _____.

   (A) a virus
   (B) a bacterium
   (C) a fungus
   (D) an animal

10. The science that studies all of the products of genes is _____.

   (A) genetics
   (B) genomics
   (C) proteomics
   (D) genetic engineering

Free-Response

1. You have just been hired as a crime scene investigator and you're at the scene of your first big case. Your team has found several drops of blood and your job is to try to figure out to whom they belong. What will you do when you get back to the lab?

2. The Marfan syndrome is a dominant genetic disorder of the connective tissue. It is caused by an abnormality of a large gene on chromosome 15. The connective tissue that it affects is fibrillin, found throughout the body, at the joints between bones, holding the retina in place at the back of the eye, and providing strength and support to the aorta and other blood vessels. Is there a place for gene therapy in the treatment plan for a Marfan's patient? If so, is there an advantage to somatic cell or germ line gene therapy in the Marfan syndrome?

3. There are people and organizations who feel strongly that introducing any genetically modified organisms (bacteria, plant, or animal) is dangerous and wrong. Others feel that introducing "improved" members of a species is helpful and can improve life for that species and for humans. What explanations and examples might exist for both sides of this argument?

Answer Explanations

Multiple-Choice

1. **D** A breeder wants to figure out if a dog is homozygous for a dominant trait and will therefore pass on that trait to 100 percent of its offspring. The breeder can cross that dog with a homozygous recessive female dog. This test is called a test cross.

2. **C** In order to isolate a specific selected gene, it can be cut out by a protein called a restriction enzyme.

3. **B** A mechanism for separating pieces of DNA by fragment size is called gel electrophoresis.

4. **A** The driving force that pulls DNA fragments across a gel plate in question #3 is a positive electrode. The negative electrode behind the wells repels the DNA fragments. Osmosis and endocytosis are both forms of cellular transport, not of movement across a gelatin plate.

5. **C** Comparing the restriction map from a known source with one from an unknown source is known as DNA fingerprinting. Comparative restriction, genetic engineering, and PCR are not about comparing DNA from known to unknown DNA sources.

6. **D** A form of genetic engineering that develops bacteria that can consume and breakdown hazardous material and oil spills is called bioremediation.

7. **A** Genetically modified drugs including hormones, clotting factors, and vaccines are all types of biopharmaceuticals.

8. **B** When the Human Genome Project was completed, scientists were surprised to find that the human genome included 23,000 genes.

9. **A** The first gene ever sequenced belonged to a virus. A bacterium, a fungus, and an animal all have much more complex genomes that that of a virus.

10. **C** The science that studies all of the products of genes is about proteins and is called proteomics.

Free-Response Suggestions

1. **You have just been hired as a crime scene investigator and you're at the scene of your first big case. Your team has found several drops of blood and your job is to try to figure out to whom they belong. What will you do when you get back to the lab?**

   First, the small amounts of blood need to be amplified, or their volume of DNA increased. This can be done by PCR, polymerase chain reaction, where small volumes of DNA can be duplicated in several hours to form millions of copies. Next, you'll want to run a gel electrophoresis using DNA samples from a victim and any other people who may have been at the scene. A comparison of restriction maps will provide a DNA fingerprint of possible perpetrators of the crime.

2. **The Marfan syndrome is a dominant genetic disorder of the connective tissue. It is caused by an abnormality of a large gene on chromosome 15. The connective tissue that it affects is fibrillin, found throughout the body, at the joints between bones, holding the retina in place at the back of the eye, and providing strength and support to the aorta and other blood vessels. Is there a place for gene therapy in the treatment plan for a Marfan's patient? If so, is there an advantage to somatic cell or germ line gene therapy in the Marfan syndrome?**

   Gene therapy may be part of a future treatment plan for a Marfan's patient, but probably not yet. The Marfan gene is large, with many possible sites for mutations and the protein produced by this gene is found throughout the body. Somatic cell gene therapy would only work on specific sites in the body and so would probably not be as effective as germ line therapy. Germ line gene therapy could potentially cause the cells of the entire body to make fibrillin correctly and would prevent the disorder from being passed on to future generations. Unfortunately for Marfan's patients, germ line therapy still has medical, technical, and legal obstacles to overcome before it becomes commonly available.

3. **There are people and organizations who feel strongly that introducing any genetically modified organisms (bacteria, plant, or animal) is dangerous and wrong. Others feel that introducing "improved" members of a species is helpful and can improve life for that species and for humans. What explanations and examples might exist for both sides of this argument?**

   People who believe that introduction of GM organisms is dangerous and wrong are generally concerned that there is insufficient evidence of the safety and benefit of these modified organisms. There are people concerned that pest-resistant crops might contain chemicals that could damage healthy people. On the other side is the argument that there are starving children and adults around the world who cannot get necessary nutrients from locally grown food and that we can now grow GM food locally that is going to give them those missing nutrients. Also, if GM food products can reduce the need for fertilizers and pesticides, the negative environmental impact of agriculture can be reduced.

CHAPTER 9

Energy

Outline for Organizing Your Notes

I. Energy Exchange with the Environment

II. Photosynthesis
  A. Light-Dependent Reactions
  B. Light-Independent Reactions

III. Aerobic Cellular Respiration
  A. Glycolysis
  B. Kreb's Cycle
  C. Oxidative Phosphorylation

IV. Anaerobic Cellular Respiration
  A. Alcoholic Fermentation
  B. Lactic Acid Fermentation

V. Energy Exchange
  A. ADP
  B. ATP
  C. Energy Storage Macromolecules

What's the big deal about living things and energy? People talk about energy all the time regarding transportation and gasoline prices, fuel availability and heating or cooling of our homes. Athletes talk about energy in terms of giving their bodies the fuel they need to perform at maximum efficiency. How does energy impact your life? Perhaps you're into fitness or maybe you're interested in how to stay better focused during a day of school or an evening of studying. Even the simple acts of blinking your eyes, swallowing, sitting up, and reading these words requires the use of energy. Whenever we talk about bodies using energy, we're referring to how the cells in that body receive, metabolize, and use energy.

Cells can't go out to get a meal or a candy bar for some quick energy. They need a way to get nourishment and energy quickly. They need to get their energy in a useable and manageable form. Most living things use carbohydrates as their main energy source. This biological molecule is manufactured by autotrophs, or producers, during photosynthesis and is used by these autotrophs and by heterotrophs during cellular respiration.

Energy Exchanges with the Environment

The basic photosynthesis–cellular respiration cycle is the foundation of energy for almost every life form on Earth today. During photosynthesis, sunlight's energy converts water and carbon dioxide, both continually entering the plant from the surrounding soil, water, and the air, into sugar and oxygen. Sugar and oxygen are then used, during cellular respiration, as food and fuel by both the autotrophs that made them and by the heterotrophs that take them in. Photosynthesis occurs in three segments, the first two are **light-dependent reactions** which are chemical energy storing phases that only happen during sunlight, and the third is the **light-independent reactions** that can happen whether or not sunlight is available. Both light-dependent and light-independent processes generate the production of carbohydrates. While they do not require the presence of sunlight, the light-independent reactions do require the products of the light reactions and are, therefore, indirectly dependent on sunlight. Cellular respiration takes the sugar and oxygen produced during photosynthesis and turns them into stored energy for the cell during the processes of **glycolysis**, the **Kreb's cycle**, and **oxidative phosphorylation**.

Photosynthesis

Energy from the sun reaches the Earth's surface as waves of light. This light hits the surface of photosynthetic organisms, plants, as well as some algae and cyanobacteria. Some light is reflected off the surface of the organism and some is absorbed. In the case of green plants which contain chlorophyll, green wavelengths are reflected and all other visible wavelengths (red, orange, most yellow, most blue, and violet) are absorbed. Different types of chlorophyll and other pigment-containing compounds are present in photosynthetic cells and absorb different wavelengths of light. In prokaryotes these compounds are present in the cytoplasm. In eukaryotes, these light-absorbing compounds and the process of photosynthesis occur in organelles called plastids. Green plastids are called chlorplasts.

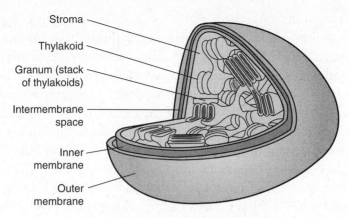

Major Structures of the Chloroplast

When light waves are absorbed into the chloroplasts, the light energy is temporarily stored in structures called **thylakoid** membranes that are stacked into structures called **grana**. This stored light energy initiates a series of events called the light-dependent reactions. The light energy triggers **photolysis**, the splitting of a water molecule into separate oxygen and hydrogen molecules, and it releases an electron. At this point, the reaction is no longer driven by light energy, it is driven by chemical energy. The oxygen is released from the plant to enter the atmosphere. The loose electron is accepted into a series of electron carriers, the **electron transport chain**, as it passes from one to the next, energy is lost. This energy is used to generate two energy-storing activities. It provides the energy to bond an inorganic phosphate to a stable molecule called adenosine diphosphate (ADP) to produce the energy-carrying molecule adenosine triphosphate (ATP). Energy for cellular reactions is held in the bond of this third phosphate to ADP. The other energy-storing activity generated by the electron carriers is the addition of the available hydrogen to a molecule of NADP, making it NADPH.

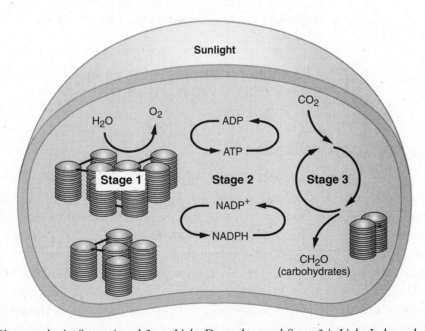

Photosynthesis: Stages 1 and 2 are Light Dependent and Stage 3 is Light Independent

The second set of photosynthetic reactions are called the light-independent reactions. These reactions occur in the **stroma**, or the fluid of the chloroplast that surrounds the grana. As the name indicates, these reactions occur regardless of the presence of sunlight. These reactions are considered to be carbon fixing because they incorporate carbon dioxide into a carbohydrate, an organic molecule. They are also reactions that require the products of the earlier, light-dependent reaction in order to proceed. During the repeated cycles that make up this set of reactions, carbon dioxide from the atmosphere interacts with the ATP and NADPH generated during the light-dependent reactions to produce glucose. Energy is released when the bond attaching the third phosphate to an ATP molecule is broken. The extra electron carried by NADPH also helps to drive this glucose-producing cycle when the hydrogen breaks away to leave the molecule NADP.

Photosynthesis Chart

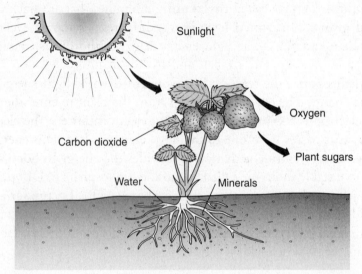

Carbon dioxide enters the leaves through stomata (tiny holes) in the leaves.

The Formula for Photosynthesis is:
$$6H_2O + 6CO_2 \rightarrow C_6H_{12}O_6 + 6O_2$$

Aerobic Cellular Respiration

Almost all autotrophs (producers) and heterotrophs (consumers), and even decomposers, use the process of cellular respiration to utilize carbohydrates to make and store energy. They use the oxygen given off during photosynthesis to assist them in metabolizing sugars. **Aerobic respiration** is a three-step process that breaks down sugar in the presence of oxygen to produce ATP.

The first step, glycolysis, involves splitting a sugar in half, into two, three-carbon molecules called pyruvates. The name of this step describes what happens; "glyco" means sugar and "lysis" means to break apart. Glycolysis occurs in the cytoplasm and produces four molecules of ATP. The process of glycolysis uses up two molecules of ATP and ultimately generates two available high-energy molecules each of ATP and NADH. The next step occurs in the mitochondria and is called the Kreb's cycle, or the Citric Acid cycle. During the Kreb's cycle, pyruvate is transported inside the mitochondria, carbon dioxide is removed, and the three-carbon pyruvate becomes a two-carbon molecule called acetyl-CoA. The Kreb's cycle ultimately generates high-energy molecules of ATP (two molecules), NADH (triggering the production of eight molecules of ATP), and $FADH_2$ (yields two molecules of ATP). During oxidative phosphorylation, the electron transport chain (ETC) embedded in the inner mitochondrial membrane allows energy to be released through the reduction of NADH to NAD+ and $FADH_2$ to FAD+. Water is released as a by-product of this process. As the energy is released, it is recaptured and stored as ATP; each molecule of NADH+ contributes three ATP molecules and each molecule of $FADH_2$ contributes two molecules of ATP. There are a total of 40 ATP molecules produced but since four of them are used up in the process of respiration, the final energy outcome of aerobic respiration is 36 available molecules of ATP.

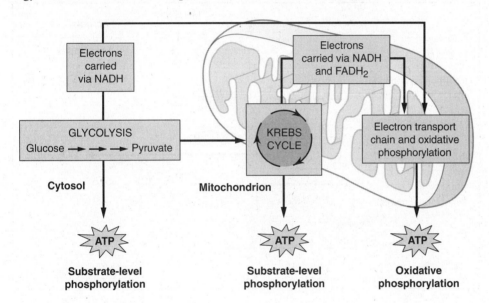

Anaerobic Respiration

Cellular respiration that occurs in the absence of oxygen is called **anaerobic respiration**. There are a number of anaerobic pathways that generate energy although none of them are as effective as aerobic respiration. Many bacteria, and other microbes, live in an oxygen-free environment and they use other molecules than sugar to get energy. Some of these energy-producing pathways include denitrification and carbonate, nitrogen, and sulfate reduction. Each of these are mechanisms that use available molecules to move electrons and to generate energy in the form of ATP.

Usually when we discuss anaerobic respiration, we are referring to the use of carbohydrates to generate energy without using oxygen to move the electrons. There are two kinds of carbohydrate-based anaerobic respiration, both of which begin with glycolysis and occur in the cytoplasm. They are both fermentation processes and each yields 10 percent of the ATP generated by aerobic respiration; four molecules of ATP produced, two used to fuel the reactions, and two molecules ultimately available as stored energy. In yeast, alcohol fermentation begins with the splitting of a glucose molecule and results in the production of ethanol, carbon dioxide, and ATP. In humans, anaerobic respiration also begins with glucose but, in the absence of oxygen, lactic acid fermentation occurs and produces lactic acid and ATP.

In humans, anaerobic respiration occurs when muscles experience intense exercise. Have you ever exercised so much that you felt that you couldn't go on? Did you drop to the ground at that point? Probably not. Your body has a backup system so that when you run out of available oxygen, you can still continue to function. Now, instead of entering the Kreb's cycle, pyruvate is converted through a series of steps into the final product, lactic acid. The lactic acid sits in your muscles, it crystallizes and when these crystals are squeezed between your muscle fibers, they cause pain. So, remember that time you kept going after you thought you couldn't go on? Were your muscles sore the next day? Probably!

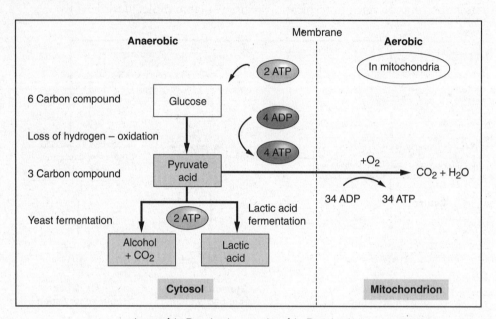

Anaerobic Respiration vs. Aerobic Respiration

Energy Exchange

Living cells use the molecule ATP as their unit of "energy currency." Adenosine triphosphate, ATP, is a molecule made of adenosine (the sugar - ribose + the nucleotide - adenine) with three phosphate groups (PO_4) bonded to it. The energy that is available in ATP is stored in the high-energy bond that holds the third phosphate group onto the molecule of ADP. Adenosine diphosphate is a stable molecule composed of adenine with two phosphate groups attached. The mechanisms of cellular respiration trigger the bonding of the third phosphate group, inorganic phosphate or P_i, onto the ADP to form ATP. When the high-energy bond holding the third phosphate group onto ATP is broken, ADP and P_i are left behind along with enough energy to fuel cell-sized reactions.

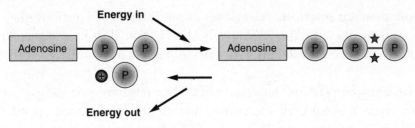

$$ADP + \text{inorganic } P \rightarrow ATP$$

ATP can be stored for a short time in an organism that has excess energy available, such as a plant during the summer or an animal that eats a lot. This energy needs to be used relatively quickly, though, because the function of ATP is short-term energy storage. If the ATP sits for too long, it will be used to produce more stable molecules, through polymerization, for stable long-term energy storage. Long-term energy storage in plants is usually maintained in polysaccharide and starch molecules. Extra energy is usually stored in animals in the form of fats, which are stored in adipose tissue.

Important Terms to Remember

Aerobic respiration: Cellular respiration in the presence of oxygen; this is a much more efficient process than anaerobic respiration in that it ultimately produces 36 available molecules of ATP.

Anaerobic respiration: Cellular respiration in the absence of oxygen; this type of respiration yields much less energy than aerobic respiration, only two molecules of ATP. Common forms include alcoholic fermentation and lactic acid fermentation.

Electron transport chain: A series of electron carriers that move energy from the splitting of a water molecule to the bonds of ATP and NADPH during the light-dependent reactions in photosynthesis.

Glycolysis: The first step of cellular respiration (aerobic and anaerobic); this step splits six-carbon glucose into two three-carbon molecules called pyruvate.

Grana: (singular: **Granum**) Stacked thylakoid membranes inside the chloroplast, the site of light-dependent reactions of photosynthesis.

Kreb's cycle: In the presence of oxygen, the 3-C pyruvate made during glycolysis is split into 2-C acetyl-CoA, CO_2 is released and ATP, NADH, and $FADH_2$ are made available to generate energy.

Light-dependent reactions: First set of photosynthetic reactions; they require sunlight to trigger photolysis; products of these reactions are O_2, ATP, and NADPH.

Light-independent reactions: Second set of photosynthetic reactions; they do not require sunlight; they use the ATP and NADPH formed in light-dependent reactions to drive a cycle of reactions with carbon dioxide to produce glucose.

Oxidative phosphorylation: Final stage of aerobic respiration; involves the electron transport chain moving available energy into the third phosphate bond of ATP. Final yield is 36 molecules of ATP.

Photolysis: The splitting of water early in photosynthesis; it releases oxygen into the atmosphere and makes available hydrogen and electrons to form the energy-providing molecules, ATP and NADPH.

Stroma: The fluid of the chloroplast that surrounds the grana; the site of the light-independent reactions.

Thylakoid membranes: Internal membranes of a chloroplast; they are the sites for storage of energy from sunlight.

Practice Questions

Multiple-Choice

1. Which biological macromolecule provides most of the energy for day–to-day life activities in living organisms?

   (A) proteins
   (B) lipids
   (C) carbohydrates
   (D) nucleic acids

2. What is the overall chemical formula for photosynthesis?

   (A) $6CO_2 + 6H_2O \rightarrow C_6H_{12}O_6 + 6O_2$
   (B) $6C_6H_{12}O_6 + 6O_2 \rightarrow 6CO_2 + 6H_2O$
   (C) $O_2 + H_2O \rightarrow C_6H_{12}O_6 + CO_2$
   (D) $C_6H_{12}O_6 + O_2 \rightarrow CO_2 + H_2O$

3. In prokaryotes, light-absorbing compounds like chlorophyll are found in the _____.

   (A) chloroplasts
   (B) cytoplasm
   (C) mitochondria
   (D) ribosomes

4. When plant cells absorb light waves, the light energy is first stored in the
   _____, a structure in the _____.

   (A) cristae; mitochondria
   (B) nucleolus; nucleus
   (C) stem; plant
   (D) thylakoid; chloroplast

5. The biggest difference between the initial reactions of photosynthesis and the
   second set of reactions is that _____.

   (A) The first reactions require direct interaction with sunlight and the
       second set doesn't require direct interaction with sunlight.
   (B) The first reactions don't require direct interaction with sunlight and the
       second set do require direct interaction with sunlight.
   (C) The first reactions use ATP for energy and the second set uses ADP for
       energy.
   (D) The first reactions don't make or store any energy and the second set of
       reactions do make and store energy.

6. Organisms that use the products of photosynthesis include _____.

   (A) plants
   (B) animals
   (C) protists
   (D) all of the above

7. During the Kreb's cycle, one high-energy molecule generates more energy
   storage than any others, this molecule is _____.

   (A) ADP
   (B) ATP
   (C) NADH
   (D) FADH$_2$

8. What is the overall chemical formula for cellular respiration?

   (A) $6CO_2 + 6H_2O \rightarrow 6C_6H_{12}O_6 + 6O_2$
   (B) $C_6H_{12}O_6 + 6O_2 \rightarrow 6CO_2 + 6H_2O$
   (C) $O_2 + H_2O \rightarrow C_6H_{12}O_6 + CO_2$
   (D) $C_6H_{12}O_6 + O_2 \rightarrow CO_2 + H_2O$

9. The final energy production of aerobic respiration is a total of _____ molecules of ATP, with _____ molecules of ATP available for use by the cell.

   (A) 40; 40
   (B) 40; 36
   (C) 32; 22
   (D) 4; 2

10. The final energy production of anaerobic respiration is a total of _____ molecules of ATP, with _____ molecules of ATP available for use by the cell.

    (A) 40; 40
    (B) 40; 36
    (C) 32; 22
    (D) 4; 2

Free-Response

1. Photosynthesis and cellular respiration are considered to be two halves of the same cycle. Explain why this is the case.

2. In a food chain with blades of grass, or microscopic phytoplankton, as the main autotrophic food source, the top heterotrophs are often very large, such as elephants, lions, and whales. How can this be the case? Explain how these food chains can work.

3. ATP is considered the "universal energy exchange" of the cell. Why? Explain.

Answer Explanations

Multiple-Choice

1. **C** Carbohydrates are the biological macromolecule which provides most of the energy for day-to-day life activities in living organisms. Proteins provide structure, hormones, enzymes, and messaging. Lipids provide insulation and long-term energy storage. Nucleic acids provide hereditary information for the organism.

2. **A** The overall chemical formula for photosynthesis is $6CO_2 + 6H_2O \rightarrow C_6H_{12}O_6 + 6O_2$.

3. **B** In prokaryotes, light-absorbing compounds like chlorophyll are found in the cytoplasm. Prokaryotes do not have chloroplasts or mitochondria. They do have ribosomes but they are used for protein synthesis, not light absorption.

4. **D** When plant cells absorb light waves, the light energy is first stored in the thylakoid, a structure in the chloroplast.

5. **A** The biggest difference between the initial reactions of photosynthesis and the second set of reactions is that the first reactions require sunlight and the second doesn't require sunlight. Both use ATP for energy and both make and store energy.

6. **D** Organisms that use the products of photosynthesis include all of the following: plants, animals, and protists.

7. **C** During the Kreb's cycle, one high-energy molecule generates more energy storage than any others; this molecule is NADH which ultimately triggers production of eight molecules of ATP. $FADH_2$ ultimately triggers production of two molecules of ATP. ADP is not a high-energy molecule.

8. **B** The overall chemical formula for cellular respiration is $C_6H_{12}O_6 + 6O_2 \rightarrow 6CO_2 + 6H_2O$.

9. **B** The final energy production of aerobic respiration is a total of 40 molecules of ATP, with 36 molecules of ATP available for use by the cell.

10. **D** The final energy production of anaerobic respiration is a total of four molecules of ATP, with two molecules of ATP available for use by the cell.

Free-Response Suggestions

1. **Photosynthesis and cellular respiration are considered to be two halves of the same cycle. Explain why this is the case.**

   Photosynthesis and cellular respiration are considered to be two halves of the same cycle. There are a few reasons for this; essentially they continuously exchange the same molecules; the formula for photosynthesis is $6CO_2 + 6H_2O \rightarrow C_6H_{12}O_6 + 6O_2$, and the general formula for cellular respiration is $6C_6H_{12}O_6 + 6O_2 \rightarrow CO_2 + 6H_2O$. So, the products of one keep being recycled as the reactants of the other. Photosynthesis uses energy, which comes from sunlight and cellular respiration stores energy in the form of ATP.

2. **In a food chain with blades of grass, or microscopic phytoplankton, as the main autotrophic food source, the top heterotrophs are often very large, such as elephants, lions, and whales. How can this be the case? Explain how these food chains can work.**

   In a food chain with blades of grass, or microscopic phytoplankton, as the main autotrophic food source; the top heterotrophs are often very large, such as elephants, lions, and whales. The autotrophs in any ecosystem are the most plentiful. Although blades of grass and phytoplankton are small, there is a huge abundance of them. The autotrophs provide a lot more of the biomass, or biological nutrients, than heterotrophs. Many large heterotrophs spend most of their waking hours finding food and eating it, in order to get enough nutrients to keep their bodies running.

3. **ATP is considered the "universal energy exchange" of the cell. Why? Explain.**

ATP is considered the "universal energy exchange" of the cell. ATP is the chemical that is used to fuel all of your cellular activities. It releases energy in just the right amount to support these activities without leftovers, or wasted energy. In this respect, using ATP is like spending a $1 bill instead of a $100 bill to buy a candy bar; it's the right amount for cellular use. ATP is also like an ATM card. You can use what you need and then recharge it and use it over and over.

CHAPTER 10

Classification

Outline for Organizing Your Notes

I.   Classification
    A. Binomial Nomenclature

II. Taxonomy

III. The Domains or Kingdoms
    A. Bacteria
       1. Archaebacteria
       2. Eubacteria
    B. Protista
       1. Animal-like Protists
       2. Plant-like Protists
       3. Fungus-like Protists
    C. Fungi
    D. Plantae
       1. Non-vascular (mosses)
       2. Vascular
          a. Ferns
          b. Gymnosperms (pine and fir trees)
          c. Angiosperms (flowering and fruiting plants)

    E. Animalia
       1. Invertebrates
          a. Porifera (sponges)
          b. Cnidaria (jellyfish)
          c. Platyhelminthes (flatworms)
          d. Nematoda (roundworms)
          e. Annelida (segmented worms)
          f. Arthropoda (insects, arachnids, butterflies, and crustaceans)
          g. Mollusca (clams, snails, squid, and octopus)
          h. Echinodermata (sea stars/starfish and sea anemones)
       2. Chordates
          a. Jawless Fish
          b. Cartilagenous Fish
          c. Bony Fish
          d. Amphibians
          e. Reptiles
          f. Aves (birds)
          g. Mammals

IV. Biodiversity

Say your cousin goes into the Peace Corps and is giving away all the stuff she doesn't need. She has given you her collection of hundreds of CD's, DVD's, and video games. They're all in boxes, so you buy new shelves for your room to store them. How will you arrange these items on your shelves? Will you separate the CD's, DVD's, and games? Within each of these groups will you organize based on alphabetical order, artist, genre, or some other categories? If you don't organize them at all, what will happen?

Let's consider other systems for organizing things in our lives. How about the grocery store or the library? Both places keep a huge supply of diverse materials that need to be accessed without too much hassle. If they weren't well organized, it would be time consuming and difficult to find the book you want at the library or that can of tuna at the grocery store. How about your own everyday organizational process? Do you organize your computer by making folders and grouping related materials in the same folder? The alternative would be to save files and then just search through your existing files for the one you want. This works when you've saved just a few files, but once you've saved a couple hundred files, it can get pretty time consuming and tedious to search out the one you want. It's the same story for anything of which you have a lot; collections, DVD's, computer games, baseball cards, even your school notebooks, need to be organized for quick access to the information you want.

Taxonomy

Have you ever wondered just how many organisms are alive in the world? Have you ever thought about how often new species are being discovered or how often other species are becoming extinct? What about some of those organisms that seem so much alike yet don't belong to the same groups, or those that seem so different but are grouped together? And, why do we need all these long, strange-sounding names and groupings anyway?

Taxonomy is the science of classification. Biological classification involves categorizing living things into groups, or taxa, that are related to each other and organizing these groups into a clear hierarchy. The job of classifying taxa belongs to scientists called taxonomists. Biological taxonomy groups all living things into kingdoms, which are then separated into phyla (or separated into divisions within the plant kingdom), which are subdivided into class, order, family, genus, and species.

The science of naming living things was first developed by Aristotle (384–322 B.C.). Aristotle grouped animals based on whether they lived primarily in the water, on land, or in the air. He grouped plants based essentially on their size; small plants were considered herbs, midsized plants were named shrubs, and large plants were called trees. This system existed for about 2,000 years.

Binomial Nomenclature

In 1735, Carolus Linneaus (1707–1778) published *Systema Naturae* in which he explained a hierarchal system to organize and classify life forms. Linneaus established a classification hierarchy including kingdom, class, order, genus, and species. He also developed the two-level naming system called **binomial nomenclature**.

Using binomial nomenclature, all living things are given a scientific name made up of their genus and species. The scientific name is written in a very specific way, with the genus name beginning in a capitalized letter and the species name beginning in a lowercase letter. The entire scientific name is always either underlined or italicized. Rather than rewriting the genus name many times when we're discussing different species of the same genus we can simply abbreviate the genus to its first letter, for example *E. coli* is the abbreviation for *Escherichia coli*. Similarly, we can write about different or multiple species of the same genus by just writing sp. (singular) or spp. (plural). This abbreviation represents species; for example, *Plasmodium* sp. indicates any individual species of *Plasmodium* and *Plasmodium spp.* represents multiple species of this malaria-producing parasite.

The Domains or Kingdoms

Since Linnaeus developed his naming system in 1735, scientific advances have altered our understanding of life forms. Linnaeus recognized three kingdoms in nature; animal, vegetable, and mineral. In 1866, advances in microscopy resulted in a change; the kingdoms being recognized were all alive (plant and animal) and the third kingdom included the mostly microscopic kingdom, protista. In the 1930's, the development of microscopes allowed scientific distinction to be made between prokaryotes and eukaryotes. Fungi were not accepted as a unique kingdom, separate from the plants, until 1969.

Current taxonomy is based upon a system originated by Carl Woese (born 1928), who separated the prokaryotes into the domain bacteria and kingdom eubacteria (the common or true bacteria) and the domain archaea and kingdom archaebacteria (the ancient bacteria). Woese grouped all other like forms into the domain eukarya which includes the kingdoms protista, fungi, plantae, and animalia. Woese used an organism's structures and functions as well as their molecular and evolutionary information to place it into the naming system correctly. This system is called the three-domain system, or the six-kingdom system, and is the taxonomic system most commonly used in America today. The classification systems used today include the following categories, in order from least specific to most specific: kingdom, phylum, class, order, family, genus, and species.

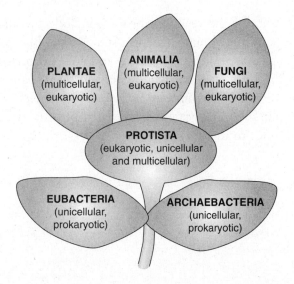

Kingdom Archaebacteria

The ancient bacteria, or **archaebacteria**, are prokaryotic organisms that either obtain their nutrients from extreme environments or just use more extreme metabolic pathways than the eubacteria. They are simple, single-celled organisms, without a nucleus or complex organelles. Archaebacteria are known as extremophiles, thriving in areas many other organisms would not survive; hot springs, salt lakes, volcanic beds, in swamps, and in the methane-filled digestive systems of ruminating animals. Archaebacteria are able to metabolize and use molecules that are toxic or, at least unusable, to most other life forms.

Kingdom Eubacteria

True bacteria, or **eubacteria**, are the single-celled prokaryotes that we encounter most commonly. Again, because we know that they are prokaryotes, we know that they are simple, single-celled organisms without a nucleus or complex organelles. Species of eubacteria are usually one of three shapes: rod, sphere, or spiral. Eubacteria are all around us and inside us. Many bacteria are helpful to humans, some inhabiting our digestive systems and helping us to digest food and make the vitamins we need. Some bacteria help us to stay healthy by keeping harmful bacteria from growing. Some bacteria can be harmful to humans. Diseases caused by bacterial infections include: tuberculosis, pneumonia, leprosy, tetanus, cholera, typhoid fever, syphilis, and diphtheria.

Kingdom Protista

The classification **protista** refers to a diverse group of organisms that may be microscopic or visible; they are eukaryotic and therefore, have a nucleus and complex organelles but are not particularly complex organisms. The sophistication of current classification systems makes it difficult to identify some of these organisms accurately; molecular testing may produce one classification while genetic or ultrastructure testing may indicate a different classification. The protista are a group of 30 to 40 loosely related phyla with significant common features regarding acquisition of food, mechanisms of movement (motility), reproduction, and cell coverings.

The major categories of the kingdom protista include the animal-like protists, plant-like protists, and the fungus-like protists. Animal-like protists are single-celled, are heterotrophs, are called protozoans, and are categorized by their mechanism for motility. Protists that move using cilia and flagella are called ciliates and flagellates. Ciliates include the common *Paramecium* and *Balantidium coli,* which are parasitic to human digestive systems. Flagellated protists include *Trichonympha* sp., which help termites to digest wood.

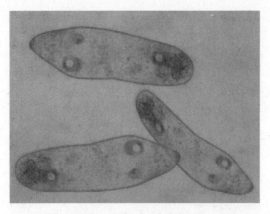

A Ciliate—*Paramecium* sp.

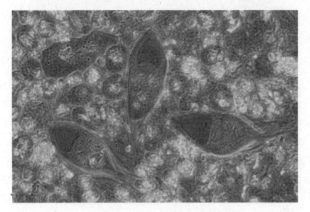

Flagellates—*Trichonympha* sp.

Protozoans that move by extending their cell membrane to surround their food are called rhizopods. The extended cell membrane is called a **pseudopodium**, or false foot. Amoebas are a common type of rhizopod. Many amoeba species are harmless to humans but *Entamoeba histlytica* is one that causes many people around the world to get sick. *E. histlytica* causes many deaths from amoebic dysentary. Another dangerous protozoan is *Plasmodium* sp. which causes malaria. *Plasmodium* belongs to a group of protozoans called sporozoans, they don't move, they tend to be parasitic, and normal body functions of their hosts move these parasites around to complete their life cycle.

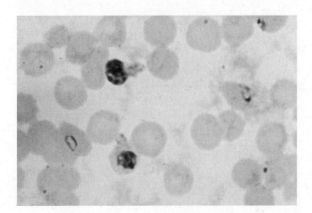

A Rhizopod—*Entamoeba histolytica*

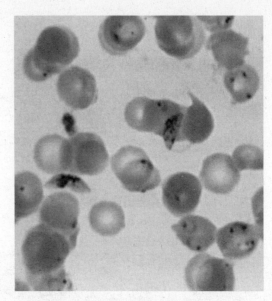

Sporozoans—*Plasmodium* sp.

Plant-like protists can be unicellular, or multicellular, and are all photosynthetic autotrophs and are aquatic (salt and fresh water). The simplest plant-like protist is *Euglena* sp., which is single-celled and is both photosynthetic and motile. Other plant-like protists include golden algae, green algae, brown algae, and red algae. Golden algae, also known as diatoms, are unicellular organisms found in both fresh and salt water. Golden algae have silica-based cell walls that make them look like glass ornaments. When diatoms die, their remaining glasslike walls become part of the sand nearby.

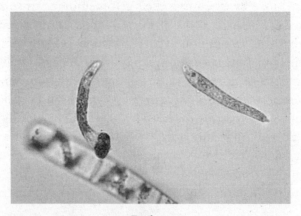

Euglena

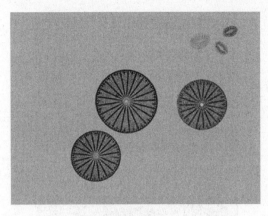

Diatoms

Green algae may be unicellular or multicellular. It is common to fresh water but can be found in salt water, too. Green algae is commonly seen as green growth in fish tanks and is closely related to the plant kingdom. Brown algae can grow to be more than 100 feet long. It is found in mostly marine, or salt water, environments but can survive in fresh water. Most seaweed, or kelp, is made up of brown algae. Red algae is also found in mostly salt water environments but can survive in fresh water. It can be found in deeper water than any of the other algae because its red pigments absorb the wavelengths that penetrate deepest into the ocean.

The fungus-like protists include slime molds, water molds, and downy mildew. Fungus-like protists are heterotrophs but have cell walls. All of them reproduce using spores, are mobile at some age in their lives, and live in moist environments. Water molds and downy mildews live on vegetation, look like fuzzy threads and can destroy entire crops. If you've ever walked deep in a forest, you may have seen a fallen tree with areas covered in slimy, brightly colored ooze. That was a slime mold. Slime molds aid in the decomposition process.

Kingdom Fungi

The kingdom **Fungi** is made up of a diverse group of eukaryotic organisms; they can be unicellular or multicellular, and live in very different habitats, some common and some quite extreme. All fungi share certain characteristics including their heterotrophic mechanism for obtaining nutrients and their stationary lifestyle. Fungi excrete enzymes to surrounding organic waste or decomposing matter to initiate or enhance the rate of decay. Fungi get their nutrients from this decomposing material. The kingdom of fungi is divided into four major categories; Ascomycota, Basidiomycota, Zygomycota, and Deuteromycota. The Ascomycota includes yeast, truffles, powdery mildews, and some blue-green and black molds. Basidiomycota includes mushrooms, puffballs, and shelf fungus. Most molds, including black bread mold, belong to the phylum Zygomycota. The fungi that don't fit into other phyla are grouped into the phylum Deuteromycota. Deuteromycota includes the fungus that produces penicillin as well as the fungi that cause athlete's foot and ringworm.

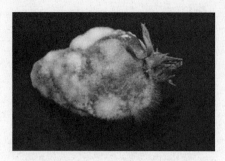

Zygomycota on Strawberry

Basidiomycota: Mushrooms

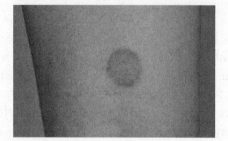

Deuteromycota: Ringworm

Kingdom Plantae

This kingdom includes all plants. The **Plantae** are multicellular eukaryotes and are stationary, photosynthetic autotrophs. The major divisions of the plant kingdom are based on their mechanisms of reproduction and how they transport materials throughout their bodies. The simplest plants are mosses and liverworts, small, nonvascular (they can't transport fluids through their bodies), spore-producing organisms that live close to the ground in moist environments. Slightly more complex are the ferns, which do have a vascular, or internal transport system, and that reproduce primarily by spores rather than by seeds.

Liverworts

Ferns

Complex plants include the Gymnosperms and Angiosperms. These are the plants that we most often notice on a daily basis. **Gymnosperms** have a vascular system allowing them to grow tall and they reproduce using seeds that are "naked," or not surrounded by an ovary. Gymnosperms never have flowers. Most often gymnosperm seeds are contained in and protected by cones; plants with cones are called conifers. The leaves of conifers are usually needle or scale shaped with a thick waxy cuticle, or cover, surrounding them. Gymnosperms include the pine, fir, juniper, cedar, and spruce trees, strong trees that can survive extreme weather.

Spruce Trees

Cedar Trees

Angiosperms, flowering plants, are the most complex of all the plants. Their transport system is capable of moving water and nutrients throughout a tall tree. They produce new plants through sexual reproduction, the seeds that they make are enclosed in a fruit (an ovary), and they make colorful and fragrant flowers to attract pollinators and seed dispersers. Angiosperms include all the flowering plants like tulips, roses, and daffodils, all the vegetables and legumes, and all the fruit trees. Angiosperms provide us with a huge variety of food products as well as decorative garden plants.

Cherry Trees

Tulips

Corn

Kingdom Animalia

The kingdom **Animalia** is made up of the most complex and diverse organisms on Earth. Animals are eukaryotic, multicellular, heterotrophic, and mobile; they lack cell walls, most reproduce sexually and most have some type of body **symmetry**, or organized arrangement. The Animal kingdom is divided into two major phyla: invertebrates and chordates. The invertebrates are organisms that don't have a backbone. Chordates all have a spinal cord and most chordates also belong to the subphylum vertebrata; telling us that they have a spinal cord enclosed within a bony case, the vertebrae. Each of these classifications is organized, by taxonomists, in order from simplest to most complex.

The Invertebrata

The simplest invertebrates are the sponges, the porifera; marine animals that are asymmetrical (have no organized shape), have only cells and tissues, reproduce both asexually and sexually, and are mobile as larvae but are sessile (permanently attached to something) as adults. The cnidaria, or jellyfish, have a gelatinous, umbrella-shaped body with radial symmetry. The Cnidaria are free swimming and are found in all the oceans of the world. They have cells, tissues, and organs, but do not have organized body systems. Cnidarians are named for their stinging cells, or cnidocytes.

Sponge

Lion's Mane Jellyfish

   Several classes of worms make up the next few levels of complexity, flatworms, roundworms, and segmented worms. Flatworms, or platyhelminthes, are bilaterally symmetrical. They have a primitive brain and simple digestive and excretory systems, exchanging both food and waste through their oral opening. Flatworms exchange oxygen and carbon dioxide through their skin, and so must maintain a relatively flat body surface for gas exchange with all cells. Nematodes, or roundworms, are a very diverse group, found in salt water, fresh water, terrestrial, and parasitic environments. Nematodes have bilateral symmetry and cylindrical bodies with a unique body cavity called a pseudocoelom, in which internal organs are found and that assists in circulation of nutrients and gases. Nematodes are the simplest organisms that have a one-way, tubular digestive system with a mouth at one end and an anus at the other end. The annelids, or segmented worms, are the most complex worms. They live in all types of marine and moist terrestrial environments. Their bilaterally symmetrical bodies are divided into many small segments, each of which contains the same set of organs. Annelids have a true coelom, or body cavity; they have a brain, and a circulatory system with pumps to help move their blood. They can reproduce sexually or asexually.

Flatworm

Roundworms

Segmented Worm

The arthropoda include insects, butterflies, lobster, crabs, scorpions, and ticks. These complex organisms have bilateral symmetry, an external skeleton (called an exoskeleton), segmented bodies, and jointed appendages. They have a circulatory system but unlike most other organisms, their blood is mostly loose in the body (open circulation). Arthropods have brains and sensory organs. They are very diverse and are one of only two groups of the animal kingdom that thrive in both dry and marine environments. The mollusca include snails, slugs, clams, squid, and octopus. Some species of mollusca, the octopus, squid, and cuttlefish, have complex brains with sophisticated nervous systems and sensory organs. Many mollusks have shells or mantles, a rasping (or grating) tongue with chitinous (similar to keratin that makes our fingernails) teeth, and a broad, muscular foot. Echinodermata are a unique group including the sea stars, brittle stars, sand dollars, and sea anemones. Echinoderm means spiny skin. These animals all live in salt water environments, have radial symmetry, and have an internal system of water-filled canals called a water vascular system.

Insect

Snail

Starfish

Phylum Chordata

Animals that belong to the phylum chordata have all, at some point in their development, had a notochord, a dorsal nerve cord, and pharyngeal gill slits. These are all structures that by adulthood, in most chordates, have developed into more sophisticated organs, have been modified, or have been lost. Most chordates belong to the subphylum vertebrata (recent taxonomic thought is that this subphylum should be renamed craniata) this grouping includes all chordates with a spinal column or backbone. The vertebrata includes jawless fish, cartilaginous fish, bony fish, amphibians, reptiles, birds, and mammals.

The class agnatha includes the jawless fish, lampreys, and hagfish. These fish have simple digestive systems, skin without scales, are **ectotherms** (cold-blooded), they do not have paired appendages, and they do retain their notochord as adults. Lampreys are parasitic, attaching to, and feeding off of, fish and ocean-dwelling mammals. Hagfish feed off dead marine animals and fend off their own predators by excreting large amounts of thick mucus. Sharks and rays belong to the class of cartilaginous fish or chondrichthyes. These animals are ectothermic, have a cartilaginous skeleton, paired fins, and toothlike scales that feel like sandpaper. The bony fish belong to a class called osteichthyes, a large and diverse group that includes ocean sunfish, tuna, trout, salmon, barracuda, and goldfish. These fish live in salt and fresh water environments, are ectothermic, and they have gills and scales.

Lamprey

Basking Shark

Ocean Sunfish

Amphibians, or members of the class amphibia, include salamanders, frogs, and toads. They are ectothermic, and many transition (undergo **metamorphosis**) from a water-dwelling and water breathing juvenile into a mature air breathing four-legged adult. Most amphibians must lay their eggs under water and have smooth, moist skin. Reptiles, members of the class reptilia, include snakes, lizards, crocodiles, alligators, and turtles. They have scaly, dry skin, are ectotherms, they have lungs, and a three-chambered heart, and can lay their eggs on land. Birds belong to the class aves. They are endothermic (warm-blooded), winged, and **bipedal** (walk on two feet). They lay hard shelled eggs, have a four-chambered heart, a beak, and feathers.

Golden Toad

Leopard Gecko

Goldfinch

Mammals belong to the class mammalia, a diverse and complex group of species. All mammals are endothermic, have sweat glands and hair, specialized teeth, a four-chambered heart, three middle ear bones used for hearing, and females produce milk in mammary glands. Mammals have a complex brain with a sophisticated neocortex. There are three divisions of the class mammalia, the monotremes, the marsupials, and the placentals. Monotremes are egg laying mammals and include the platypus and the echidna. All other mammals give birth to live young. The marsupials are pouched mammals whose young are born in an extremely immature and underdeveloped state; they climb into the mother's pouch, attach to a nipple, and grow to a point when they are mature enough to emerge. Marsupials include kangaroos and wallabies, koalas, and opossums. Placental mammals are the most common and diverse of the mammalian subclasses. Placental mammals have internal gestation with the fetus connected to the mothers blood supply by a placenta. These mammals give birth to live young, who live independently of their mother's body. Many young placental mammals do require long-term parental care to reach maturity. Placental mammals include mice, bats, bears, dogs, elephants, and humans.

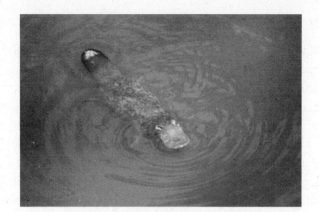

Platypus

Koalas

Polar Bear

Biodiversity

So, how many different kinds of living things live on Earth today? Currently, the numbers range between 1.4 and 2.0 million species, depending on your source. Estimates exist that there are between 5 and 100 million species that exist and/or have existed on Earth. How many new species are discovered and how many become extinct each year? Well, answering this question is tough for a lot of reasons, including: the lack of a complete, centralized database to identify these species; the discoveries and losses of entirely new communities in isolated and extreme regions, the density of populations of microscopic species in the soil and in the oceans, and simply the vast number of living things that do exist on Earth. One thing that most scientists do agree upon is that each year, far too many species are becoming extinct or endangered.

Important Terms to Remember

Animalia: Kingdom of multicellular, heterotrophic, mobile eukaryotes.

Archaebacteria: Kingdom of ancient, or extreme, bacteria.

Binomial nomenclature: Two-name naming system; used in biology to identify an organism's genus and species.

Bipedal: An organism that walks on two feet.

Ectotherm: Cold-blooded; body temperature is controlled by external environment.

Endotherm: Warm-blooded; body temperature is controlled by internal environment.

Eubacteria: Kingdom of the common and true bacteria.

Fungi: Kingdom of multicellular, heterotrophic, stationary eukaryotes that get nutrients by secreting enzymes that digest organic material for fungal absorption.

Metamorphosis: Transformation of an organism from larval to adult stages involving different physical appearances in the different stages.

Plantae: Kingdom of multicellular, autotrophic, stationary eukaryotes.

Protista: Kingdom of simple eukaryotes. They may be autotrophs or heterotrophs, unicellular or multicellular, stationary, or mobile.

Pseudopodium: Selective extensions of the cell membrane and cytoplasm that act as false feet.

Symmetry: An organism's body arrangement; asymmetrical organisms have no organized arrangement; radial symmetry indicates a circular body arrangement with similar body sections arranged like a pie; bilateral symmetry indicates a linear body arrangement with two similar sides.

Taxonomy: The science of classification; groups of organisms are taxa. These are grouped from least specific to most specific by kingdom, phylum, class, order, family, genus, and species.

Practice Questions

Multiple-Choice

1. Which of the following is the correct order of levels of classification from most to least specific?

   (A) Kingdom → Phylum → Class → Family → Order → Genus → Species
   (B) Kingdom → Phylum → Order → Genus → Family → Class → Species
   (C) Species → Genus → Order → Class → Family → Phylum → Kingdom
   (D) Species → Genus → Family → Order → Class → Phylum → Kingdom

2. There are many types of scientists that participated in your Amazon expedition. The scientists whose job it is to classify all the living things you encounter are called _____.

   (A) categorists
   (B) classifists
   (C) scientific classification specialists
   (D) taxonomists

3. The science of naming organisms today takes into account all the following except _____.

   (A) evolutionary history
   (B) structure
   (C) where they live
   (D) function

4. You're looking through a powerful microscope at a single-celled organism that doesn't appear to have a nucleus or any obvious organelles. What are you most likely observing?

   (A) a prokaryotic cell
   (B) a protist cell
   (C) a tiny animal cell
   (D) a eukaryotic cell

5. You observe a moving, single-celled organism through your microscope. It has a nucleus and it appears to have many organelles including a flagellum. You can be pretty certain that it is a(n) _____.

   (A) algae
   (B) protozoan
   (C) eubacterium
   (D) fungus

6. Your little brother was playing in the sandbox at the playground. Now he has a raised, red, circular infection on the skin of his arm. Your mom says it's ringworm. His infection is caused by a _____.

   (A) worm
   (B) eubacteria
   (C) protist
   (D) fungus

7. Which of the following characteristics of sponges might make you question whether sponges are animals?

   (A) They are heterotrophs.
   (B) They are multicellular.
   (C) Their bodies are not composed of complex organ systems.
   (D) They are stationary as adults.

8. It was raining last night and when you took your dog for a walk this morning you saw both earthworms and a slug moving across the sidewalk. How are they different?

   (A) Earthworm has no legs; slug has legs.
   (B) Earthworm has a segmented muscular body; slug has nonsegmented muscular body.
   (C) Earthworm has no exoskeleton; slug has an exoskeleton.
   (D) Earthworm belongs to mollusca; slug belongs to nematoda.

9. Most animals are ectotherms, or are cold-blooded. Which of the following are endotherms?

   (A) birds
   (B) fish
   (C) snakes
   (D) frogs

10. A platypus is an unusual mammal. Which of the following make it unique?

   (A) It lays eggs.
   (B) It is cold-blooded.
   (C) It has no hair.
   (D) Platypus females do not produce milk for their offspring.

Free-Response

1. How does taxonomy help scientists around the world? Explain.

2. Thoroughly explain the following chart:

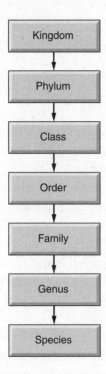

3. Is there a connection between taxonomy and evolution? Explain.

Answer Explanations

Multiple-Choice

1. **D** The order of levels of classification from most specific to least specific is Species, Genus, Family, Order, Class, Phylum, and Kingdom. Species is most specific because it contains the organisms that are the most alike.

2. **D** Taxonomists are the scientists who find the best ways to organize living things. They organize taxa into hierarchal groups with similar structures, functions, and molecular and evolutionary patterns.

3. **C** Taxonomists today consider similarity of structures, functions, and molecular and evolutionary patterns in establishing groupings of related organisms. Where they live ceased to be a consideration when scientists began to understand that structures, functions, and molecular and evolutionary patterns were more significant indicators of a relationship than whether something lives in the water, on land, or in the air.

4. **A** You are likely to be looking at a prokaryotic cell. You know this because it does not have a nucleus or obvious organelles. Most eukaryotic cells do have an obvious nucleus and organelles.

5. **B** Of the available choices, a living, single-celled organism with organelles, including a flagellum, is most likely to be a protozoan, or animal-like protist. Algae, eubacteria, and fungi are unlikely to have a flagellum.

6. **D** Ringworm is a fungal infection caused by fungi from the phylum deuteromycota.

7. **D** Sponges are stationary as adults; this is unusual for animals because animals are heterotrophs that get their nutrients by finding and eating either autotrophs or other heterotrophs. Adult sponges get their food as it drifts past them in the water; they don't have to move as adults. Larval sponges do pursue food as they are free swimming.

8. **B** Both the earthworm and the slug have muscular bodies; the earthworm's body, though, is segmented while the slug's is not. Neither of them has legs or an exoskeleton. The earthworm belongs to the annelida and the slug belongs to the mollusca.

9. **A** Birds and mammals are the only endothermic, or warm-blooded, classes of animals. The classes of fish, amphibians (like the frog), and reptiles (including snakes) are all cold-blooded, or ectothermic.

10. **A** A platypus is a monotreme, a mammal that lays eggs. There are no ectothermic mammals. All mammals have hair and female mammals are all designed to produce milk for their young.

Free-Response Suggestions

1. **How does taxonomy help scientists around the world? Explain.**

   Taxonomy allows scientists around the world to have a common language when they are referring to different organisms they may be encountering or working with. Taxonomy allows for a higher probability of reproducibility in scientific experiments because scientists around the world know that they are referring to the exact same species of organism. Taxonomy also provides a strong organizational basis for investigating life forms. The establishment of scientifically supported relationships between species allows for more extensive and thorough scientific inquiries.

2. **Thoroughly explain the following chart:**

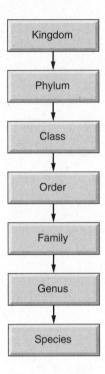

The chart presents a sequence of levels of organization in all living things: Kingdom, Phylum, Class, Order, Family, Genus, and Species. The highest level, Kingdom, includes the most types and the biggest variety of organisms. Each successive step down contains fewer types of organisms which are increasingly more similar to each other. A genus has distinct types of organisms that are clearly related to each other but that are still different enough to make it unlikely that they would interact in nature or breed with each other. A species is a group of organisms that are closely related to each other; members of a species can breed together and produce fertile offspring.

3. **Is there a connection between taxonomy and evolution? Explain.**

There is a strong and significant connection between taxonomy and evolution. One way to look at this relationship is to follow each of the kingdoms from their simplest to their most complex members and then to identify the simplest to the most complex kingdoms. If these sequences of complexity are compared to an evolutionary history of life on Earth, the two tend to parallel each other. Another way to view the relationship between evolution and taxonomy is that as species evolve, they continue to share some traits with their common ancestors and with other species that also evolved from the same common ancestors. These traits will undoubtedly be a consideration as the related species are classified.

Complex Mechanisms of Transport

CHAPTER **11**

Outline for Organizing Your Notes

I. Simple Transport
    A. Diffusion
    B. Small Organisms with All Their Cells in Close Proximity to a Watery Environment

II. Transport in Fungi
    A. Heterotrophs
    B. Secretion of Enzymes
    C. Absorption of Nutrients
    D. Rhizomes

III. Transport in Plants
    A. Autotrophs
    B. Xylem
    C. Phloem

IV. Transport in Animals
    A. Heterotrophs
    B. Simple Diffusion
    C. Open Circulation
    D. Closed Circulation

All cells need to be in a fluid environment to stay alive. This is relatively easy to do if you are a microscopic or single-celled organism; if your source of water dries up, you just build yourself an external container (called a spore) and become dormant until moisture reappears. Or perhaps you stay in the area where your life began and, as long as it doesn't dry up completely, you may be fine. What if, though, you are multicellular, mobile, or really large? How do all the living cells in your body stay surrounded by a fluid environment? How do all the living cells in a mouse or an elephant, a six-inch dandelion or a 300-foot redwood tree, or even a small mushroom maintain their external and internal environments to ensure cellular hydration?

Why is a water-based environment so important anyway? The overriding reason is that the inside of every living cell is filled with cytoplasm; this is a water-based matrix that holds organelles in place and where most metabolic reactions of the cell occur. Cytoplasm has amino acids and proteins, simple sugars and complex carbohydrates, and different types of lipids all mixed in, but the fact remains that water is the main ingredient in cytoplasm. Water is also a reactant or a product in most biochemical reactions. The cell membrane recognizes water and is selectively permeable, allowing water in and out of the cell when other molecules can't pass through. If a cell is in a hypertonic environment, water leaves the cell and it may become limp or even implode. If a cell is in a hypotonic environment, water will flood into the cell, causing it to swell and possibly even explode. Water has a role in enzymatic reactions and in the electron transport chain and formation of ATP. Water's polar interactions with the phospholipid bilayer create an environment where the cell membrane stays intact. Without water, cell membranes and cytoplasm could not exist. Without cell membranes and cytoplasm, cells could not exist and without cells, life would not exist.

Simple Transport

The simplest and smallest organisms tend to live in moist environments or, if they don't, they've developed adaptations that allow them to store water or encapsulate themselves until water becomes available. Many prokaryotes and protists use their habitats to provide water and nutrients through simple diffusion and osmosis. As long as they are surrounded by an isotonic solution, they can maintain dynamic equilibrium and their cells can function normally. These organisms depend on their watery environment to maintain their cell membranes and cytoplasm, to bring them oxygen and nutrients, and to get rid of their wastes. All of these activities are performed through direct interaction with their surroundings.

Transport in Fungi

There are a number of types of fungi; the trait they all share is their general mechanism for gaining nutrients. Fungi are heterotrophs, so they don't produce their own food, like plants do. Fungi are also stationary, so they can't move to catch their food. You might think that this would make it difficult for fungi to obtain nutrients and survive, but because they've developed some unique adaptations, they do just fine.

Fungi get their nutrients from other organisms. They excrete digestive enzymes that will break down the complex molecules that are accessible and transform them into smaller organic molecules that the fungus can absorb through its cell walls. Fungi are the only kingdom that obtains its nutrients by **absorption**, by digesting external nutrients and then absorbing them. To understand the transport process in fungi, you should know that the bodies of fungi are made up of entangled multitudes of fibers called **mycelia** (singular: mycelium). The filaments, or fibers, that make up the mycelium are called **hyphae** (singular: hypha). Fungi can reproduce sexually or asexually by producing spores, or by budding in yeast. Fungal spores are blown through the air and if they land and germinate on a surface they can digest, the fungi will thrive by growing and extending their hyphae into their "host." Fungi secrete enzymes that are specific to the once-living material that they've grown into, decomposing the surface they've landed on, and using its components as nutrients.

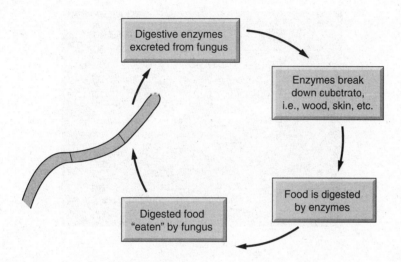

So, as far as transporting nutrients is concerned, fungi are not all that different from humans. Fungi encounter a source of nutrients; they can quickly absorb simple organic molecules like glucose, but most of the food they find is in the form of complex biological molecules. They digest those molecules externally and then absorb them. We humans take in our food, digest it internally, and then absorb nutrients through the lining of our intestines. Food, to a fungus, could be bread, rotting fruit, an animal carcass, hair, or a fallen tree. Many species of fungi only produce enzymes that will digest a specific kind of food. So if you see a mushroom growing on a tree that fell over in the woods, you can be pretty certain that it makes enzymes that will break down cellulose and lignin, two large organic molecules that the fungus will encounter, digest, and absorb from the dead tree. This same fungus is not likely to be able to digest the polysaccharides found in an old loaf of bread.

Penicillium sp.

Shelf Mushrooms

Bread Mold

Transport in Plants

The movement of water and nutrients through plants is very different from their transport in fungi. Tiny plants, like mosses, stay small and live in a moist environment so that their cells can be in constant contact with water and the nutrients that it carries. As plants get bigger and taller, it becomes ever more important to have structural adaptations that allow water to move around and to carry nutrients and other essential molecules to all the living cells of the organism. Remember, not only do water and some minerals need to get up from the soil to the highest branches and leaves, but the sugars produced during photosynthesis need to be transported down through plants to nourish the stem and root cells.

Imagine yourself standing at a second- or third-story-level window at your school or home. You're thirsty and there's a delicious, cold drink of water available outside on the ground below your window. You just happen to have a 30-foot straw stretching from the drink to your mouth. Do you think you could draw water that distance up the straw? Have you ever considered that a tree does this all the time? The major factors that support transport, against the forces of gravity, in taller plants like angiosperms and gymnosperms, are structural adaptations like those that allow **capillary action** and **transpiration** and the characteristics of water, such as **cohesion** and **adhesion**.

Transport of water and dissolved mineral ions begins via osmosis into the root hair cells extending from the plants roots into the soil. If active transport is necessary to provide adequate concentrations of the minerals, the plant will use ATP to fuel this important exchange. Water and minerals enter the root system and travel up through the plant through vessels called **xylem**. The cells of xylem form a wide, tough, thick-walled tube traveling from the roots, through the stem or trunk, and up to the highest leaves. The lives of xylem cells begins like other plant cells as immature cells called **parenchymal cells**, but as they mature, they begin to accumulate a tough, fibrous material called lignin that makes the cell walls very rigid and ultimately kills them. When xylem cells die they lose their cytoplasm and end walls but remain in place, strengthening both the transport system and the support system for keeping the plant standing tall.

Capillary action and transpiration help to move water and its dissolved minerals upward, in spite of the downward pull of gravity. Capillary action is the tendency of liquids to rise up into a thin tube, based on an attraction between solid and liquid materials. The properties of water contribute to the capillary action seen in xylem. Cohesion is the property that holds two water molecules together, you see this when water maintains the shape of the hose after leaving it when you water your lawn. Adhesion is the property that attracts and holds water to a solid surface, like the beads of water on your skin when you climb out of a pool. Transpiration is the movement of water out of the leaves of a plant, similar to evaporation; water is released creating a suction that pulls water from the roots upward through the xylem toward the leaves.

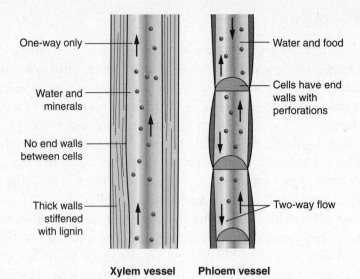

One-way only

Water and minerals

No end walls between cells

Thick walls stiffened with lignin

Water and food

Cells have end walls with perforations

Two-way flow

Xylem vessel　　**Phloem vessel**

Transport through Xylem and Phloem

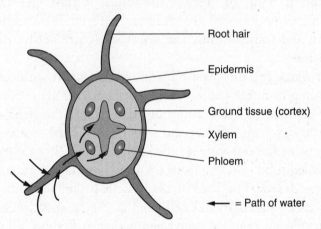

Root hair

Epidermis

Ground tissue (cortex)

Xylem

Phloem

← = Path of water

Cross Section of a Root

Nutrients move throughout the plant in a different kind of tubular cells called **phloem**. Phloem is a tubular system that flows both up and down carrying the products of photosynthesis from the leaves to cells everywhere in the plant. Phloem cells are alive and serve only in transport, not in adding strength or support for the plant. Unlike xylem, phloem is comprised of two types of cells, sieve cells and companion cells. Sieve cells form the tube that sap moves through; they are porous and allow for easy movement by the dissolved sugars they carry. Sieve cells do not contain nuclei or most other organelles; in order to stay alive, each sieve cell has a companion cell that takes care of the life functions for itself and for its associated sieve cell.

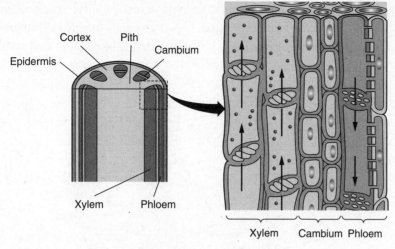

Longitudinal Section of a Stem

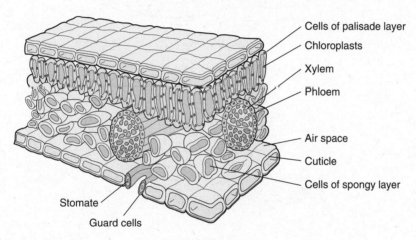

Cross Section Through a Leaf

Transport in Animals

The simplest members of the animal kingdom obtain their nutrients through diffusion. Organisms, including sponges, jellyfish, and flatworms, exchange oxygen and carbon dioxide, nutrients, and waste by keeping all of their cells in close proximity to the watery environment in which they live. Roundworms have a pseudocoelom, or false body cavity, which allows water to flow past its internal cells. This extended surface area provides opportunities for cells that aren't in contact with the external environment to exchange nutrients, waste, oxygen, and carbon dioxide.

Transport in more complex animals is called circulation. While there are varying levels of complexity depending on the species, there are only two major kinds of circulation in animals; transport through open and closed circulatory systems. Open circulation happens when blood is pushed into the body cavity by the heart. Once the blood is pumped out of the heart, it is not contained in true blood vessels. Instead, blood and other body fluids enter a **hemocoel** which is a closed cavity with a series of gaps and spaces between the organs. Blood gases, nutrients, and waste are exchanged with surrounding cells by diffusion. Blood returns to the heart by moving past the gills in mollusks and through the **spiracles** into the **tracheal system** of insects, where O_2 and CO_2 are exchanged. Open circulation is found in arthropods, such as insects and crustaceans, and in some mollusks, like clams and snails.

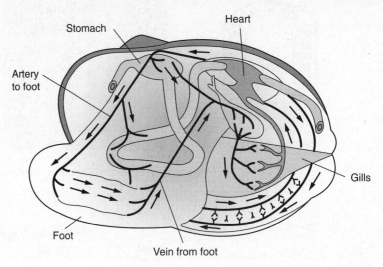

Open Circulation in a Mollusk

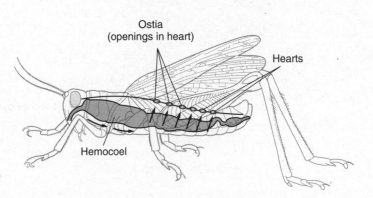

Open Circulation in an Arthropod

All of the vertebrates (fish, amphibians, reptiles, birds, and mammals) as well as the most complex mollusks (the squid and octopus) have a closed circulatory system. In a healthy closed circulatory system, the blood is contained inside the heart or vessels all the time. The heart acts as a pump and the vessels are the tubes that move the blood throughout the body. Blood vessels include thick-walled arteries to pump blood away from the heart, thick-walled veins to carry blood back to the heart, and thin-walled capillaries between the two. The capillaries are the sites for the exchange of O_2 and CO_2 and the exchange of nutrients with waste. In the bodies of organisms with closed circulation, the capillaries are in direct contact with all living cells.

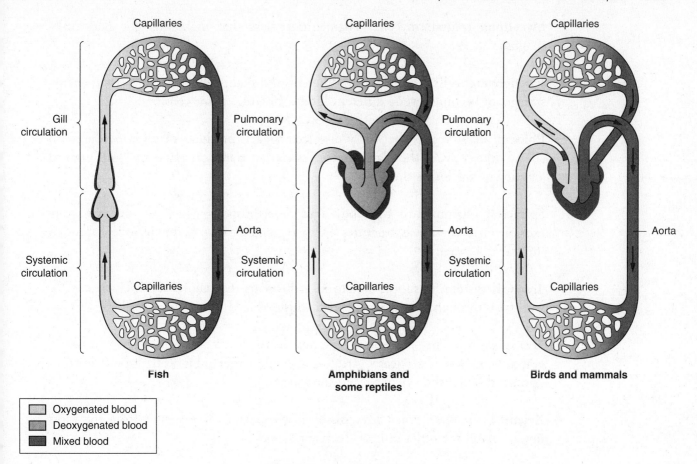

Circulatory paths in vertebrates

Important Terms to Remember

Absorption: The mechanism of acquiring nutrients in fungi; the fungus secretes digestive enzymes and then takes in the digested material through diffusion into its hyphae.

Adhesion: The property that attracts and holds water to a solid surface, like beads of water sticking to the walls of the shower.

Capillary action: The tendency of liquids to rise up into a thin tube based on an attraction between solid and liquid materials.

Cohesion: The property that holds two water molecules together; this property allows water to maintain the shape of the faucet opening after it leaves the faucet.

Hemocoel: A series of spaces between the organs in the body cavity of an organism with open circulation, essentially arthropods and mollusks.

Hypha: (plural: hyphae) Single fungal filament.

Mycelium: (plural: mycelia) Filamentous mass that makes up the body of a fungus.

Parenchymal cells: Young and undifferentiated plant cells that can serve in energy storage, or become several different kinds of mature plant cells.

Phloem: Plant vessels that carry dissolved sugars (products of photosynthesis) to all the other cells of the plant. These vessels can transport materials both upward and downward inside the plant.

Spiracles: Openings in the body wall of arthropods where O_2 and CO_2 are exchanged; these surface structures allow gas exchange with the environment through the tracheal system.

Tracheal system: A simple system in arthropods that allows the exchange of O_2 with CO_2 through a network of small, chitin-ringed tubes called tracheae.

Transpiration: The movement of water out of the leaves of a plant, similar to evaporation; water is released creating a suction that pulls water from the roots upward through the xylem toward the leaves.

Xylem: Plant vessels that carry dissolved water and mineral ions from the roots upward to all the other cells of the plant.

Practice Questions

Multiple-Choice

1. Water provides many services to cells, which of the following does water not do for a bacterium?

   (A) brings nutrients
   (B) gets rid of wastes
   (C) triggers development
   (D) provides oxygen

2. When many species of bacteria are exposed to drought conditions how do they survive?

   (A) They move to a new area that has water.
   (B) They group together and wait for water to return.
   (C) They form a protective spore and become dormant until water reappears.
   (D) They don't survive; they dry up and die without water.

3. Why do plants need to have separate transport vessels for moving materials up and down the plant?

   (A) Because water and minerals need to travel up the plant and carbohydrates need to be transported up and down.
   (B) Because carbohydrates need to travel up the plant and water and minerals need to be transported down.
   (C) Because it's easy to transport carbohydrates, water, and minerals down the plant using gravity, but transporting carbohydrates, water, and minerals up the plant requires work.
   (D) The plant doesn't need separate transport vessels.

4. If large molecules or ions need to use energy for their active transport into or out of a cell, that energy will come directly from a molecule of _____.

   (A) protein
   (B) carbohydrate
   (C) ADP
   (D) ATP

5. Vessels that carry materials up the plant from the roots to the leaves are called _____ and they transport _____.

   (A) xylem; water and minerals
   (B) xylem; carbohydrates
   (C) phloem; water and minerals
   (D) phloem; carbohydrates

6. Transport cells which become support cells in plants that do that job best only after they are dead are called _____ and they die because they accumulate a tough, fibrous material called _____.

   (A) xylem; parenchyma
   (B) xylem; lignin
   (C) phloem; parenchyma
   (D) phloem; lignin

7. Some simple animals have a false body cavity which gives them extra internal surface area for water to exchange nutrients, gases, and wastes with the cells. This structure is called a _____.

   (A) pseudocoelom
   (B) pseudocavity
   (C) hemocoel
   (D) spiracle

8. Oxygen enters the bodies of invertebrates with open circulation through _____.

   (A) pores and tracheoles
   (B) arteries and veins
   (C) gills and spiracles
   (D) hemocoels and ostia

9. Many animals have a closed circulatory system, including all of the following except _____.

   (A) birds and octopus
   (B) snakes and mice
   (C) squid and humans
   (D) snails and scorpions

10. In animals with closed circulation, the blood vessels that allow the exchange of O_2 with CO_2 and of nutrients with waste are called _____.

   (A) veins
   (B) arteries
   (C) capillaries
   (D) all of the above

Free-Response

1. Why is a water-based environment important to all cells?

2. How do plants ensure that water is available to all of their living cells?

3. How do animals ensure that water is available to all of their living cells?

Answer Explanations

Multiple-Choice

1. **C** Water does not trigger development. It does bring nutrients, gets rid of wastes, and provides oxygen to a bacterial cell.

2. **C** Bacteria exposed to drought conditions survive by forming a protective spore and becoming dormant until water reappears. They can't make a decision to move to a new area that has water. They would die if they grouped together to wait for water to return. Without protective spores they probably won't survive in the absence of water.

3. **A** Plants need to have separate transport vessels for moving materials up and down the plant because water and minerals need to travel up the plant and carbohydrates need to be transported throughout the plant.

4. **D** If large molecules or ions need to use energy for their active transport into or out of a cell, that energy will come directly from a molecule of ATP.

5. **A** Vessels that carry materials up the plant from the roots to the leaves are called xylem and they transport water and minerals.

6. **B** Transport cells in plants that do their job best only after they are dead are called xylem and they die because they accumulate a tough, fibrous material called lignin.

7. **A** Some simple animals have a false body cavity which gives them extra internal surface area for water to exchange nutrients, gases, and wastes with the cells. This structure is called a pseudocoelom.

8. **C** Oxygen enters the bodies of invertebrates with open circulation through gills and spiracles.

9. **D** Many animals have a closed circulatory system, including all of those listed except snails and scorpions.

10. **C** In animals with closed circulation, the blood vessels that allow the exchange of O_2 with CO_2 and of nutrients with waste are called capillaries.

Free-Response Suggestions

1. **Why is a water-based environment important to all cells?**

   Water availability is important to all cells for many reasons. One reason is that the inside of every living cell is filled with cytoplasm; this is a water-based solution that holds organelles in place and where many metabolic reactions of the cell occur. Water is a reactant or a product in most biochemical reactions. The cell membrane recognizes water and is selectively permeable, allowing water in and out of the cell when other molecules can't pass through. Water has a role in enzymatic reactions and in the electron transport chain and formation of ATP. Water's polar interactions allow the cell membrane to stay intact. Without water, cell membranes and cytoplasm could not exist and without cells life would not exist.

2. **How do plants ensure that water is available to all of their living cells?**

   Water and minerals enter the root system and travel up through the plant through vessels called xylem. The movement of water inside a plant is a result of some very important properties of water, allowing it to travel upward through the plant in spite of gravity. These properties include cohesion and adhesion, capillary action, and the ability to undergo transpiration.

3. **How do animals ensure that water is available to all of their living cells?**

Simple and small organisms have water surrounding all their cells. As animals get larger and more complex, they require a circulatory system to ensure water, or blood, availability to all cells. Most invertebrates that have a circulatory system have open circulation, which means that they have a heart that pumps blood into an open body cavity to bathe the cells. More complex organisms, octopi, squid, and all vertebrates, have closed circulation. Their heart pumps blood into a series of tubes, or arteries, that carry blood to all parts of the body. Arteries feed blood into thin-walled capillaries that allow the exchange of needed and waste materials between their water-based carrier, blood, and the cells throughout the body. After O_2 and nutrients diffuse into the cells and CO_2 and wastes diffuse out of cells, and back into the blood, they are carried back to the heart through veins.

Evolution and Natural Selection

CHAPTER 12

Outline for Organizing Your Notes

I. Evolution of Life on Earth
   A. Before Life Evolved
   B. Prokartotes
   C. Eukaryotes
      1. Protista
      2. Fungi
      3. Plantae
      4. Animalia

II. Charles Darwin
   A. Travels on the HMS Beagle
   B. Development of the Theory of Natural Selection

III. Theory of Natural Selection

IV. Evidence for Evolution
   A. Observation
      1. Peppered Moths
      2. Antibiotic Resistant Bacteria
   B. Fossil Evidence
   C. Structural Evidence
      1. Homologous Structures
      2. Analogous Structures
      3. Vestigial Structures
   D. Embryological
   E. Biochemical

V. Patterns of Evolution
   A. Divergent Evolution
      1. Adaptive Radiation
      2. Geographic Isolation
      3. Reproductive Isolation
      4. Selective Breeding
   B. Coevolution
   C. Convergent Evolution

VI. Human Evolution
   A. Mammals
   B. Primates
   C. *Hominidae*
      1. *Australopithecus* sp.
      2. *Homo habilis*
      3. *Homo neanderthalensis*
      4. *Homo erectus*
      5. *Homo sapiens*

Think of all the different kinds of living things that you can. At first you might be thinking of people, dogs, cats, horses, ferrets, maybe cows, sheep, and goats. Maybe you'll think of fish, frogs, snakes, worms, and spiders. Then you remember that living things also include plants, fungi, amoeba, algae, and bacteria. There are a lot of different kinds of living things on this planet and there are many more that once thrived but are now **extinct**. Do these organisms have a stronger connection to each other than simply living on the same planet? What is the correlation between athlete's foot fungus and an oak tree? How are sharks and mildew related? What is the relationship between seaweed and an earthworm, between *E. coli* bacteria and a scorpion, or between an ear of corn and you?

While I'm sure we'd agree that they are all very different and unique, we'd also have to recognize that all living things have many characteristics in common. They all grow, develop, are internally organized, reproduce, maintain homeostasis, metabolize nutrients, and get rid of waste products. Every living thing contains nucleic acids, DNA and RNA, to direct their reproduction, whether they are producing offspring or new cells for their multicellular body. The DNA and RNA in every living thing perform the same functions and are made of the same molecules. The genes coded in our DNA dictate the sequence of amino acids to make the proteins to make every living organism on Earth. Every living thing shares DNA sequences that code for many of the same proteins; we're just assembled differently based on the DNA that we inherited from our parents, their parents, and their great-great grandparents. Our DNA as humans can be traced back to Africa about 60,000 years ago but we can also recognize genetic relationships between humans and other species, even other kingdoms, that goes back much farther in time. Study of these relationships helps us to understand the science of **evolution**.

Evolution of Life on Earth

Scientists have established that the Earth formed about 4.6 billion years ago. The atmosphere was hot and turbulent and would have been toxic to all forms of life; it was made up of methane, ammonia, and other poisonous gases. As millions of years went by, the Earth's crust cooled enough to allow continental plates to form. Chemical elements mixing in the oceans included hydrogen, oxygen, carbon, nitrogen, phosphorus, and sulfur. These elements, and more, created a **primordial soup**, a mixture of elements that were mixed violently, jolted with electricity, and heated. Lightning storms raged, pummeling the oceans with jolts of electricity. Tides crashed the ocean into shorelines and then receded, leaving tidal pools baking in the intense sunlight. The ingredients of the primordial soup formed unique bonds that eventually became the basis for the development of life on earth.

The first evidence of life appears around 3.8 billion years ago. This life would have been in the form of primitive pre-cells, using simple carbohydrates for nutrients and containing very basic proteins, lipids, and nucleic acids. These pre-cells are likely to have interacted and even colonized to form the earliest true cells, bacteria, around 3.5 billion years ago. As bacterial species evolved and diversified, some developed the ability to use energy from sunlight to make sugars to nourish themselves. Cyanobacteria (a prokaryote) and blue-green algae (a eukaryote) were actively conducting photosynthesis around two billion years ago. A by-product of this photosynthetic activity was the release of oxygen, which began to accumulate in the atmosphere approximately 1.8 billion years ago. Keep in mind that the activities described in this one short paragraph actually occurred over more than two billion years! That's TWO BILLION years of DNA being passed from one generation to the next, with mutations occurring, species changing, and new species forming that have had different and often better adaptations to the environment in which they live. Life on earth was evolving.

The last billion years have been a time of tremendous growth and diversification of species on Earth. The earliest multicellular organisms began to thrive in the oceans around 650 million years ago. By 550 million years ago, the seas were teeming with bacteria and eukaryotic organisms; both unicellular and multicellular life forms were flourishing and the first lichens were becoming established on land. Approximately 460 million years ago, red and green algae, worms, marine invertebrates, corals, and primitive fish were thriving in the world's oceans and some vascular plants began to move onto land. By 370 million years ago, the first seed plants and many species of animal had appeared on land; land-dwelling arthropods and four-legged vertebrates had begun colonizing terrestrial Earth. All of these organisms had DNA unique to their species but also had DNA linking them back to the earliest bacteria.

Around 250 million years ago, the largest mass extinction in Earth's history occurred. While its cause has not been conclusively established, it is scientifically accepted that huge numbers of species were wiped out. This extinction was particularly destructive to marine species; it is estimated that up to 96 percent of the marine species and up to 70 percent of the terrestrial vertebrate species became extinct during this event. Following this mass extinction, new and different species had the opportunity to thrive and now reptiles took the lead, beginning the "Age of the Dinosaurs." The earliest mammals appeared around 240 million years ago and the earliest angiosperms (flowering plants) became established around 145 million years ago.

Another mass extinction occurred 65 million years ago, causing the extinction of the dinosaurs and many other kinds of animals. Many life forms did survive this mass extinction event and around 30 million years ago, the environment on Earth was right for the proliferation of mammals and flowering plants. The earliest ancestors of the human race belonged to the genus *Homo*. These primates appeared around 1.8 million years ago and the first *Homo sapiens* or humans appeared about 60,000 years ago in Africa.

Charles Darwin's Work

Charles Darwin (1809–1882) wrote a book called *On the Origin of Species*, in which he described the process by which populations of living organisms evolve. This book was published in 1859 and represented many years of Darwin's observations and ideas. Darwin grew up loving nature and preferring to learn from nature itself rather than from a teacher in a classroom. He spent a lot of his time exploring and examining the world around him. In 1831, not sure of the path he was going to take in life, Darwin signed on to travel on the ship HMS. Beagle. He planned to explore nature on continents he hadn't yet visited and agreed to be a companion to the ship's captain during their anticipated two-year trip. The voyage ended up lasting five years and Darwin ended up with volumes of intricate drawings and notes along with a huge collection of biological specimens.

Back at home in England, Charles Darwin spent most of the next 25 years trying to understand and explain the relationships between the organisms he had seen on his travels. Darwin had, for example, identified very distinct and different species of finch on the Galapagos Islands. Each species was well adapted to the unique ecosystem of the island they inhabited and each had features of the finches of the Ecuadorian mainland. He identified similar distinctions between the Galapagos tortoises from the ecologically different islands. It happened time and again; Charles Darwin was seeing organisms that were different from their ancestors, and those differences reflected adaptations that helped them, and their offspring, survive more effectively in their environment.

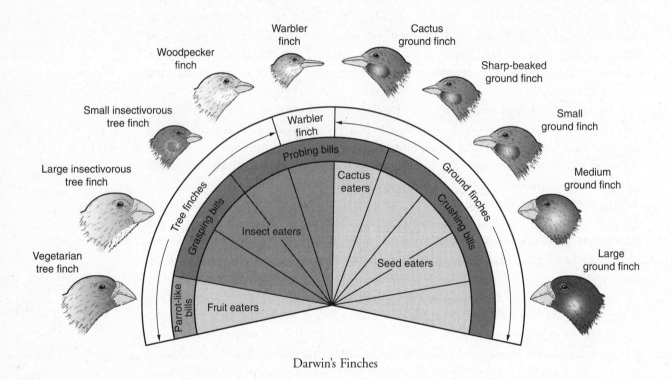

Darwin's Finches

Theory of Natural Selection

Charles Darwin was an open-minded thinker, scientist, explorer, and naturalist but his greatest contribution to science was his identification and explanation of the means by which species evolve, his **Theory of Natural Selection**.

Darwin's Theory of Natural Selection

1. Organisms overproduce offspring. Many of these offspring die before they have the opportunity to reproduce.
2. There is variation between members of a species and these variations are passed on from parents to their offspring.
3. Some organisms have variations that are better adapted to their environment than others.
4. Better adapted organisms are likely to live longer and produce more surviving offspring than less adapted organisms.
5. As beneficial variations keep more and more offspring alive, those variations become characteristic to the species.

One frequently used example of the Theory of Natural Selection is the story of the peppered moth in England, during the Industrial Revolution. In England, in the early 1800s, peppered moths were most commonly pale-colored moths with a few beige or gray spots on their wings. During this time, darker-colored peppered moths stood out more from the structures they might settle upon and were quickly eaten by predators. As new industry poured pollutants into the atmosphere, surfaces that were once clean and pale in color became dark and dirty. The dark-colored peppered moths now blended in more with their perches, while the pale-colored moths began to stand out and they became the prey. Over several decades, the population of peppered moths in England changed in color because the darker and better adapted members of the species survived to produce offspring.

At the same time that Darwin was publishing his book *On the Origin of Species*, Gregor Mendel was working in his gardens and beginning to understand the inheritance of traits. Neither Darwin nor Mendel was alive to hear how their ideas merged. In the early 1900s, scientists began exploring the chemical basis of inheritance and soon began to realize that many of the variations that Darwin recognized between members of the same species were the result of random mutations. These random mutations are a driving force in natural selection.

Evidence for Evolution

The peppered moth story is an example of observation of evolution. As time went by, the appearance of the species was modified based on natural selection. Fewer pale-colored peppered moths and many more dark-colored moths survived to pass on their genes for coloration to their offspring. Humans today, unfortunately, have frequent opportunities to observe bacterial evolution in action. Think of the last time you had a sore throat; perhaps you even went to the doctor. If the doctor saw evidence of a bacterial infection, he or she may have prescribed antibiotics for you. Many antibiotics are supposed to be taken for a ten-day course of treatment. So, let's imagine you're on day five of your ten-day prescription and your throat feels all better, almost all of the harmful bacteria have now been destroyed. The antibiotics also destroy good bacteria that help you with digestion of food, causing you to have an upset stomach. When you forget to take your pill on the sixth morning and your throat still feels fine and now your stomach feels better, you decide you're done taking the medicine. You can expect to continue feeling well for a few days, maybe even a couple weeks. Unfortunately, that small percentage of bacteria that hadn't been killed off by your five days on the antibiotic are the stronger bacteria. They may have had a mild resistance to the drugs you were taking. Now, these are the population of bacteria inhabiting your throat and they do not respond to antibiotics as well. You have evolved for yourself a strain of antibiotic-resistant bacteria.

Fossil evidence is another form of support for evolution. Fossils may be the mineralized remains of an organism, the impression (casting) of an organism that once was alive, or of something that was left by a living organism (footprints, burrows, feces). Fossils may be embedded in many substances including rock, tar, and amber. As centuries go by, soil and rock accumulate and are compressed on the surface of the Earth. In undisturbed rock, there are layers called strata, the deepest layers containing the oldest rocks and fossils. If fossils are found in undisturbed rock beds, they can be dated based on how deeply they are found in the rock, their **stratification**. The oldest fossils on record are those of early bacteria that lived 3.7 to 3.8 billion years ago. It is difficult, though, to document the fossil record through soft tissue because soft tissue decays quickly and easily so there is little fossilized record of its existence. Early fossilized evidence exists of such ancient ocean dwelling organisms as 13-foot-long nautiloids and 28-inch-long trilobites, both of which lived from 500 to 470 million years ago.

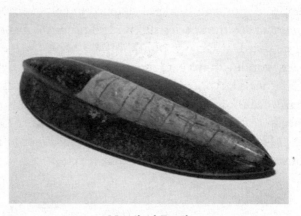

Nautiloid Fossil

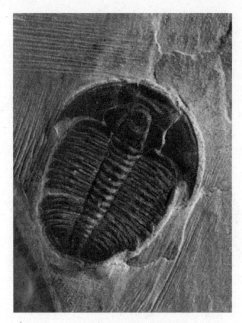

Trilobite Fossil

Homologous, **analogous**, and **vestigial structures** are all anatomical structures in diverse organisms that allow them to be compared to one another. Homologous structures are body parts that indicate common ancestry. The fact that most reptiles, birds, and mammals all have the same type of bones in their forearms indicates that they all share common ancestors. These skeletal structures may not look identical or serve the same function in each species but they are all arranged in the same sequence (humerus, radius and ulna, carpals, metacarpals, and phalanges). Analogous structures indicate separate lines of descent. Analogous structures include wings on a bird and wings on a butterfly. Both serve the same purpose, allowing their owners to fly, but they do not share internal anatomy or common ancestry. Vestigial structures are organs that have no apparent use in that species. Vestigial structures generally indicate common ancestry and are often functional structures in closely related species. Examples of vestigial structures include tail bones in humans, eyes in blind cave-dwelling salamanders, and pelvic bones in whales.

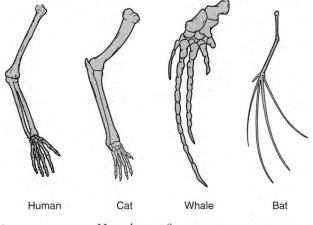

Human   Cat   Whale   Bat

Homologous Structures

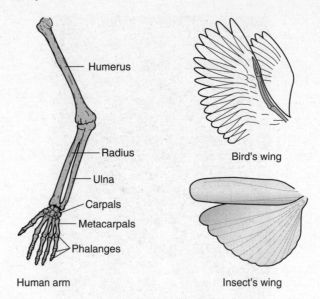

Analogous Structures

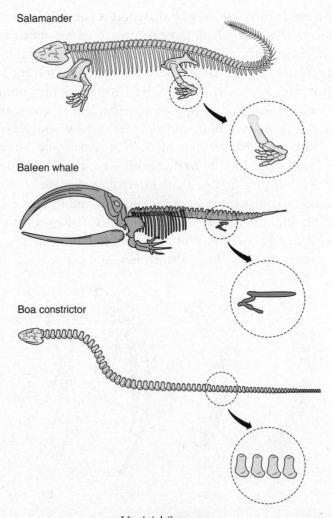

Vestigial Structures

Embryology, or developmental biology, is the study of fetal development and of the comparison of fetuses from different species. It is important to recognize that embryos of all species look different from adult members of their species. Evolutionary lines of descent may be indicated when fetuses of different species have similar appearances. While sizes and time frames may vary, there are developmental stages for most vertebrates, when fetal appearances clearly parallel one another. This correlation between species does provide some suggestion of common ancestry.

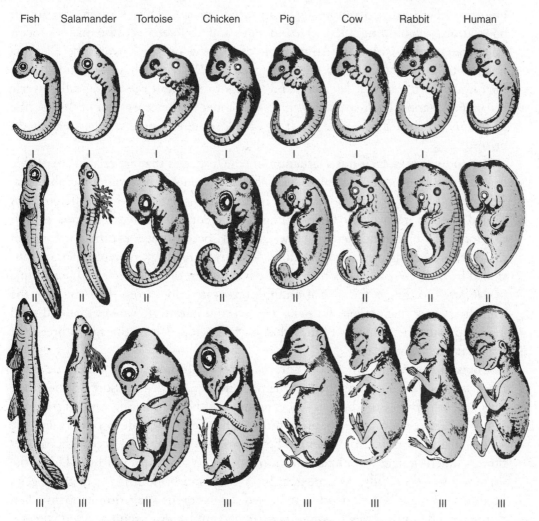

Embryological Comparisons as Evidence of Common Ancestry

Biochemical evidence of evolution can be found in similarities in the components of blood, muscle tissue, nerve tissue, and sensory tissues of different, distantly related species. Mechanisms used by all living cells for trapping and transforming energy, for building proteins, and for metabolizing nutrients are common to all species and, again, suggest common ancestral lineage. Molecular genetics is at today's core of evidence for evolution. The transcription of DNA to RNA, and then RNA's translation into specific proteins, is a process common to all living things and is considered to be strong evidence for evolutionary relationships between all living things on Earth.

Patterns of Evolution

Divergent evolution is the process of two or more related species becoming less alike over time. When things diverge, they move away from each other. An example of divergent evolution may be found in the differences in structure and function of the bones of the forearms of bats, birds, apes, humans, and the pectoral fins of a whale, all derived from the forearm structures of a common ancestor. **Adaptive radiation** is a form of divergent evolution that occurs when many unique species evolve from a single ancestral species. This is often a result of populations establishing themselves in new and different ecosystems, a process called **geographic isolation**. An example of adaptive radiation would be Darwin's finches; today, there are more than 80 species of finch that have descended from the original bird that first left mainland Ecuador to inhabit the ecologically diverse Galapagos Islands. Each species of finch is adapted to the ecosystem of the specific island on which it lives; some are fruit eaters, some eat insects, others eat cactus or seeds.

Reproductive isolation is a situation in which species that are related are unable to breed together because of physical, behavioral, genetic, or geographic obstacles. An example of reproductive isolation is temporal isolation, when different, but related, species are available for breeding at different times of year. Salmon exhibit this form of isolation; different, closely related species breed at different times of the year, preventing any cross breeding. **Selective breeding**, or reproduction controlled by humans, can play a role in the development of new species. One example of selective breeding is the speciation of domesticated dogs from their ancestor, the wolf. It is likely that ancient humans selected to domesticate wolves that had mild temperaments and other attributes that have developed into the many breeds of dog that we know today.

Coevolution occurs when two unrelated species influence each other's development as a species. Both members of the pair exert a selective pressure on the other, causing both species to change. Coevolution may be seen in the mutualistic relationship between some species of orchid that have extremely deep flowers and the moths that get nectar from them. The moths that are able to access the deeper flowers inherit longer mouthparts from their parents and these moths pollinate the deeper-flowered orchids. As time goes by, average moth mouthparts of the species become longer and flower depth is increased. **Convergent evolution** happens when distinctly different species become increasingly similar. An example of convergent evolution is the development of an opposable thumb of opossums that is only, otherwise, seen in primates. Convergent evolution is also seen in the fact that anti-freeze protein, which is made by and protects fish of the Arctic and the Antarctic, evolved independently at opposite ends of the Earth.

Human Evolution

The evolution of humans begins with the evolution of all life on Earth. We can follow the divergent evolutionary timeline that reflects the development of man by looking at the fossil record. The earliest mammals lived about 220 million years ago (mya). They were small, endothermic insectivores with milk glands and a brain with an improved capacity for complex thinking. These mammals thrived and adapted for millions of years. Then, 65 mya, one divergent group of mammals evolved into primates, characterized by a larger brain, flattened facial features, and refined hands and feet. Primates continued to adapt to their environment and that resulted in the evolution of tarsiers, monkeys, and apes in Africa 40 mya. Thirty million years ago, some primates migrated to South America and became the New World monkeys. The primates that remained in Africa ultimately evolved into the Old World monkeys and the apes. Between 7 and 15 million years ago, the great apes, or family *Hominidae*, emerged; this family included orangutans, gorillas, chimpanzees, and eventually, humans.

Between 3.7 and 2.5 million years ago, *Australopithecus* sp. developed into the earliest **bipedal** ancestors of the human race. The *australopithecines* lost their body hair, became fully bipedal, and continued to thrive in the savannahs of Africa, ultimately evolving into the earliest members of the genus *Homo*. The emergence of the *Homo habilis* 2.5 mya corresponds with the earliest appearance of tools in the fossil record. At the same time, the ancestors of today's chimpanzees and gorillas were adapting to different selective pressures and were undergoing divergent evolution from human ancestors. Evolving in Africa around 1.6 million years ago, *Homo erectus* were early ancestors of the human race who walked completely upright, had facial features that resembled modern humans, and used fire as a tool. *Homo erectus* died off around 70,000 years ago. The Neanderthals, or *Homo neanderthalensis*, lived from 300,000 to 27,000 years ago and coexisted with *H. erectus* and *Homo sapiens*.

The organisms that provide us with the oldest fossilized evidence of *Homo sapiens* lived in Ethiopia, 195,000 years ago. Evidence of death rituals, indicating strong social connections, has been found with *Homo sapiens* fossils dating back 160,000 years. Modern human cultural and social interactions gained strength and began to spread and intensify after genes related to speech developed 70,000 years ago. *Homo sapiens* migrated into Asia 50,000 years ago and into Europe and Australia 40,000 years ago. When the species *Homo floresiensis* became extinct 12,000 years ago, *Homo sapiens* became the only living species of the genus *Homo*. By the year 10,000 B.C.E. (Before Common Era), humans were farming, domesticating animals, and developing civilizations.

Important Terms to Remember

Adaptive radiation: A form of divergent evolution which occurs when many unique species evolve from a single ancestral species.

Analogous structures: Structures that serve the same purpose but do not share internal anatomy or common ancestry, indicating separate lines of descent.

Bipedal: Walking on two feet; this frees up the arms and hands for other work.

Coevolution: Occurs when two unrelated species influence each other's development as a species.

Convergent evolution: Distinctly different species become increasingly similar.

Divergent evolution: The process of two or more, related species becoming less alike over time.

Embryology: The study of fetal development and of the comparison of fetuses from different species.

Evolution: The changes that occur in a population over time.

Extinct: A species that is no longer living.

Fossil: The remains or the impression of an organism that once lived or of something that was created by a previously living organism that is embedded in rock.

Geographic isolation: When populations establish themselves in new and different ecosystems they begin to evolve to meet their local selective pressures.

Homologous structures: Body parts that indicate common ancestry. These structures may not look identical or even serve the same function in each species but do clearly originate in the same ancestor.

Primordial soup: The mixture of elements that existed in the oceans of early Earth; they were mixed violently, jolted with electricity, and heated. These elements formed unique bonds that eventually became the basis for the development of life on Earth.

Reproductive isolation: Related species that are related are unable to breed together because of physical, behavioral, genetic, geographic, or other obstacles.

Selective breeding: Reproduction controlled by humans generally to obtain offspring with favorable characteristics to the breeder.

Stratification: The levels of rock and soil created over millennia, often with fossils embedded. The deepest levels of undisturbed strata should contain the oldest fossils.

Theory of Natural Selection: Darwin's explanation of how species evolve.

Vestigial structures: Organs that have no apparent use in that species. These structures generally indicate common ancestry and are often functional in closely related species.

Practice Questions

Multiple-Choice

1. When the molecules in the early oceans churned, heated, and were jolted with electricity, their bonds were rearranged and new molecules were formed. The first development of life on Earth is linked to these newly formed molecules. What is this assortment of molecules and conditions called?

    (A) adaptive ocean
    (B) evolutionary ocean
    (C) primordial soup
    (D) primitive soup

2. The earliest life forms were bacteria that lived in the oceans. In the course of their metabolism, some of these prokaryotic organisms produced a molecule that became essential for life forms to move onto land. This molecule was _____.

    (A) water
    (B) carbon dioxide
    (C) methane
    (D) oxygen

3. Why did the dinosaurs become extinct?

    (A) Cave men killed them off.
    (B) There was a mass extinction event.
    (C) There was a severe Ice Age.
    (D) An infectious disease killed them off.

4. The mineralized impression of the burrow of a mammal that lived 15 million years ago is called a _____.

    (A) vestigial structure
    (B) strata
    (C) casting
    (D) convergent structure

5. You've been participating in an archeological dig in a newly discovered and undisturbed area. The newest artifacts are being found in the shallowest depths while the artifacts found much deeper in the earth are being dated to much earlier times. This is an example of _____.

    (A) fossilization
    (B) stratification
    (C) artifactology
    (D) human interference

6. The flightless wings of an ostrich are examples of _____.

    (A) vestigial structures
    (B) analogous structures
    (C) homologous structures
    (D) convergent evolution

7. Which of the following does not provide strong evidence of evolution?

    (A) adaptive radiation
    (B) convergent evolution
    (C) molecular genetics
    (D) developmental Biology

8. Humans practice selective breeding when they _____.

    (A) Take animals from the wild and put them into a zoo.
    (B) Hunt animals that threaten human populations.
    (C) Hunt animals that have overpopulated their ecosystem.
    (D) Control pollination of their farmed vegetables.

9. The earliest direct ancestors of the human race lived about 220 million years ago. They were _____.

    (A) mammals
    (B) warm-blooded
    (C) insect eaters
    (D) all of the above

10. The genus that is believed to be the most recent direct ancestor of the human genus, *Homo*, is _____.

    (A) *Australopithecine*
    (B) *Hominidae*
    (C) Chimpanzee
    (D) Old World monkeys

Free-Response

1. Describe the difference between saying that humans and apes descend from common ancestors and saying that humans evolved from apes.

2. Explain why nucleic acids are considered to be the common evolutionary link between humans and all other living and extinct organisms.

3. Use Darwin's Theory of Natural Selection to explain why there are more than 30,000 species of fish, 20,000 species of daisy, and more than 8,000 species of sponge living on Earth.

Answer Explanations

Multiple-Choice

1. **C** Primordial soup is the name for the assortment of molecules and conditions in the oceans that are linked to the earliest development of life on Earth.

2. **D** Oxygen is the molecule (produced by the earliest photosynthetic life forms that lived in the oceans) that, in its gaseous state, became essential for those life forms moving onto land. Gaseous and vaporized water, carbon dioxide, and methane already existed, making the Earth's atmosphere toxic to living things.

3. **B** There was a mass extinction event 65 million years ago that caused the dinosaurs to become extinct. "Cave men" could not have killed them off because nothing resembling humans existed until two to four million years ago, the dinosaurs had already been extinct for more than 60 million years. There may have been an Ice Age or an infectious disease that killed some dinosaurs but there is no fossil evidence that either would have made them all extinct at that time.

4. **C** The mineralized impression of the burrow of a mammal that lived years ago is a type of fossil called a casting.

5. **B** Stratification is when the newest fossil artifacts are found in the shallowest depths of undisturbed Earth while the artifacts found much deeper are dated to much earlier times.

6. **A** The flightless wings of an ostrich are examples of vestigial structures.

7. **B** Convergent evolution means the unrelated species evolved similar traits separately and for different reasons, it does not provide strong evidence of evolution; adaptive radiation, molecular genetics, and developmental biology do provide strong evidence of evolution.

8. **D** Humans practice selective breeding when they control pollination of their farmed vegetables. When humans take animals from the wild or hunt for them they are not conducting selective breeding.

9. **D** The earliest direct ancestors of the human race were mammals, endotherms, and insectivores.

10. **A** The genus that is believed to be the direct ancestor of the human genus, *Homo*, is *Australopithecine*. They lived 3.7 to 2.5 million years ago. The *Hominidae*, chimpanzee, and Old World monkeys all share similar ancestry but further back in time, between 7 and 15 million years ago.

Free-Response Suggestions

1. **Describe the difference between saying that humans and apes descend from common ancestors and saying that humans evolved from apes.**

   Humans and apes both live today and are examples of divergent evolution from a common ancestor. That ancestor was an early hominid that was evolved from even earlier primates and was very different from the orangutans, gorillas, chimpanzees, and humans that live today. Each species of primate has undergone unique changes based on the advantageous adaptations that gave them the best chance to thrive in their environment.

2. **Explain why nucleic acids are considered to be the common evolutionary link between humans and all other living and extinct organisms.**

   Nucleic acids are considered to be the common evolutionary link between humans and all other living and extinct organisms because the DNA that exists in every living thing is made of the same molecules as in every other living thing. All DNA works similarly in all living things and the DNA present in every living thing has been passed from generation to generation since life first began on Earth.

3. **Use Darwin's Theory of Natural Selection to explain why there are more than 30,000 species of fish, 20,000 species of daisy, and more than 8,000 species of sponge living on Earth.**

   Darwin's Theory of Natural Selection explains the huge variety of species because, in every population of a species, parents pass their traits on to their offspring. If those parents are the best adapted to their environment, they will survive longer and produce more offspring than the less-adapted parents. Offspring that are better adapted will survive in larger numbers and will live longer and produce more offspring than their less well-adapted counterparts. When this pattern continues over many generations, the populations of fish, daisies, sponges, and every other species undergo divergent evolution, and different adaptations become the norm for a whole new species of organism.

CHAPTER 13

Human Body Systems

Outline for Organizing Your Notes

I. Integumentary
  A. Skin
    1. Epidermis
    2. Dermis
    3. Hypodermis (subcutaneous tissue)
  B. Accessory Structures
    1. Glands
    2. Hair
    3. Nails
  C. Disorders
    1. Common Problems (rashes, acne, burns, blisters, lacerations, and warts)
    2. Sun Damage (Sunburn and skin cancer)
    3. Severe and Rare Disorders (ichthyosis and xeroderma pigmentosum)

II. Skeletal
  A. Bones
    1. Made of Bone, Cartilage, Blood Vessels, Fatty Tissue, Nerve Tissue, Marrow, and Connective Tissue
    2. Connected to Each Other by Ligaments
  B. Disorders
    1. Common Problems (fractures, osteoporosis, rickets, and arthritis)
    2. Severe and Rare Disorders (osteogenesis imperfecta and fibrodysplasia ossificans progressiva)

III. Muscular
  A. Muscles
    1. Three Types (skeletal, smooth, and cardiac muscle)
    2. Connected to Bones by Tendons
  B. Disorders
    1. Common Problems (cramps and strains)
    2. Severe and Rare Disorders (muscular dystrophy)

IV. Cardiovascular
  A. Circulation
    1. Heart—Acts as Pump
    2. Vessels (arteries, capillaries, and veins)—Act as Transport Tubules
    3. Blood—Carries Nutrients and Gases
  B. Disorders
    1. Heart Disease (hypertension, atherosclerosis)
    2. Heart Pain (angina pectoris)
    3. Heart Attack (myocardial infarction)
    4. Stroke

V. Respiratory
  A. Breathing
    1. Outside of the Lungs (nose mouth, trachea)
    2. Inside the Lungs (bronchi, bronchioles, alveoli)
  B. Disorders
    1. Common Disorders (bronchitis and asthma)
    2. Lung Cancer (pollution and smoking)
    3. COPD (chronic obstructive pulmonary disease)

VI. Digestive
   A. Alimentary Canal
      1. Mouth, Esophagus, Stomach, Small and Large Intestines
      2. Tube That Carries Food; Allows Nutrient Breakdown and Absorption
   B. Accessory Organs
      1. In Mouth (teeth, tongue, salivary glands)
      2. Abdominal Organs (liver and pancreas)
   C. Disorders
      1. Common Disorders (vomiting and diarrhea)
      2. Inflammatory Bowel Diseases (Crohn's disease and ulcerative colitis)

VII. Excretory
   A. Structures
      1. Kidneys (filtration of blood and production of urine)
      2. Transport (ureters and urethra)
      3. Storage (urinary bladder)
   B. Disorders
      1. Common Disorders (urinary tract infections and kidney stones)
      2. Nephritis

VIII. Nervous
   A. Structures
      1. Brain (to think and interpret, to recognize sensations and information, to store information as memory, to speak and understand language, to coordinate or movements, to anticipate and plan)
      2. Spinal Cord and Nerves (carry and transfer information)
   B. Disorders
      1. Stroke
      2. Multiple Sclerosis
      3. Alzheimer's Disease

IX. Endocrine
   A. Glands
      1. Adrenal, Pituitary, Thyroid, Parathyroid Glands, Pancreas, Ovaries, and Testes
      2. Secrete Hormones; Some Through Tubes or Ducts (exocrine glands), Some Secrete Directly into the Body Environment (endocrine glands)
   B. Disorders
      1. Diabetes (Types I & II)
      2. Reproductive System Problems, Hirsutism, Pancreatitis

X. Reproductive
   A. Structures
      1. Male (testes, epididymus, vas deferens, seminal vesicles, urethra, seminal vesicles, prostate, and the bulbourethral glands)
      2. Female (ovaries, fallopian tubes, uterus, cervix, vagina)
   B. Disorders
      1. Male (Infertility, enlarged prostate, cancer)
      2. Female (Infertility, STDs, endometriosis, cancer)

XI. Lymphatic
   A. Structures
      1. Lymphatic Fluid, Lymph Vessels, Nodes and Ducts, Red Bone Marrow, the Thymus Gland, and the Spleen.
   B. Disorders
      1. Allergies
      2. Autoimmune Diseases (lupus)

Did you ever wonder why we humans are put together the way we are? Why do we have hairs in our nose or wax in our ears? Why do we need to eat particular foods to stay healthy and how exactly does eating something make us healthier? How do the nutrients in the food we eat even get to the body parts that need them? How do we make tears and why do our knees bend in only one direction? What exactly happens that makes our skin, or bones, or any part of our bodies heal after an injury? How do our nails and hair keep growing even after our bodies have reached adult size? Why do some people get sick when others don't?

Remember that the levels of organization in living things follows this sequence: an organism is made up of organ systems, these systems are made up of organs, and organs are made up of tissues, which are made up of the smallest units of living matter, cells. Cells are made of molecules and molecules are made of atoms. To really understand how the human body works we must explore it at all these levels: cell, tissue, organ, and system. Most **anatomists**, or scientists who study how the body is put together, identify 11 major body systems that work together to make up human beings. Let's examine these systems based on levels of organization and then we can consider some of the problems that can happen when a system is damaged or isn't working correctly.

Integumentary System

The integumentary system is made up of the skin and its accessory structures; glands, hair, and nails. The skin is unique for several reasons. It is the largest organ in the body and it is also our most visible and obvious organ. The skin provides us with protection from the environment, with clues about how effectively the body is working, with temperature regulation, and with physical sensations. The skin allows for absorption and excretion, storage of blood, and the manufacture of vitamin D.

The skin is made up of two generalized layers of tissue, the dermis and epidermis. The epidermis is the outermost layer of skin and the outermost layers of epidermal cells are dead cells. These dead epidermal cells are what the world sees when they look at you. Most cells of the epidermis contain **keratin**, the structural protein found in hair and nails. Keratin makes the skin relatively waterproof. Another component of the epidermis is melanin, which gives the skin its color. The dermis is made mostly of collagen and elastic fibers. The dermis contains different types of nerve endings, hair follicles, oil, and sweat glands. Below the dermis is a subcutaneous layer made of fatty and connective tissue. The subcutaneous layer is important to the skin's function because it helps to anchor the skin and it contains the arteries and veins that supply the skin with blood.

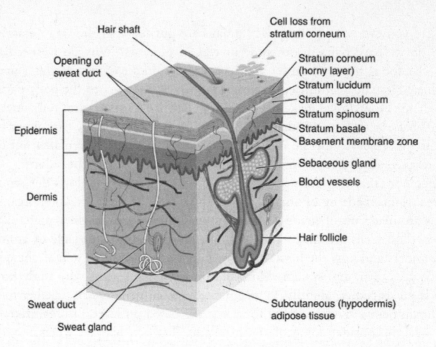

Cross Section of Skin

Common disorders and problems of the skin include rashes, acne, burns, blisters, lacerations (cuts), and warts. Sun damage to the skin and its potential result, skin cancer, are both common in the United States. Ichthyosis and xeroderma pigmentosum are both rare integumentary diseases. There are more than 20 different forms of ichthyosis, some genetic, with differing levels of severity, but all of them cause dry, thickened, flaking, or scaling skin. Xeroderma pigmentosum (XP) is a genetic disorder in which the skin cell mechanism to repair damage from ultraviolet (UV) light is inadequate. Skin cancers are very common in XP patients. In some cases, the only solution is to completely avoid UV light. Some families dealing with severe XP will completely reverse the family schedule to accommodate sleeping during the day and then perform all of their normal household activities when it's dark, and safe from UV light, outside.

Skeletal System

The skeleton of a normal adult human contains 206 bones. Each of these bones is a living organ made up of many types of tissue: bone, cartilage, blood vessels, fatty tissue, nerve tissue, marrow tissue for the manufacture of blood cells, and connective tissue. The jobs of the skeleton include movement, support, protection of internal organs, storage of minerals and triglycerides, and production of blood cells. Bones can be grouped according to their shape: long bones in the legs and arms; short bones in the wrists and ankles; flat bones like the breastbone, skull and shoulder blades; and the irregular bones like the kneecaps and vertebrae. Joints are places where bones come together. At joints, bones hold on to other bones using tissue called a **ligament**. There are hinge joints like the elbow and fingers, ball and socket joints like the hips and shoulders, pivot joints that allow us to rotate our hands and head, and fixed joints that hold two bones together without moving.

Common disorders include fractures, osteoporosis, rickets, and arthritis. Fractures, or broken bones, generally happen as a result of an injury or trauma.

Osteoporosis, the loss of bone tissue and density, causes bones to become fragile and is most common in women older than 55. Rickets is a softening and weakening of the bones as a result of a lack of Vitamin D, calcium, or phosphate. There are different types of arthritis, all of them involve pain, stiffness, and swelling at the joints. Less common are two genetic disorders, osteogenesis imperfecta (OI) and fibrodysplasia ossificans progressiva (FOP). A child with OI, or brittle bone disease, has a gene that interferes with the body's production of collagen, causing bones to break easily. Some of these children are first diagnosed after their parents are accused of abuse because of the number of broken bones seen on an X-ray. A patient with FOP has muscle, tendon, and ligament tissue which turns into bone. FOP is the only disease known that causes tissue from one body system to transform into a different type of tissue from a different system. Ultimately, a patient with FOP will develop a second skeleton which encloses the body in bone.

Muscular System

The muscular system is so closely linked to the skeletal system that they are often referred to collectively as the musculoskeletal system. There are three types of muscle in our bodies, skeletal, smooth, and cardiac. Skeletal muscle, also known as striated or voluntary muscle, is directed by our conscious choices. Skeletal muscle is attached to our skeleton, in pairs, by extensions of muscle called tendons. When one member of the muscle pair contracts, the other relaxes; these **antagonistic muscle** pairs allow all of our movements. The biceps brachialis and triceps brachialis are one pair of antagonistic muscles. If you flex at the elbow, you are contracting the biceps and relaxing the triceps. If you straighten your forearm, you relax the biceps and contract the triceps. Smooth muscle is found in the insides of our hollow and tubular organs. Smooth muscle is also known as nonstriated or involuntary muscle. Smooth muscle lines the stomach and intestines, blood vessels, uterus, urinary bladder, and airways, its contractions squeeze digesting food through the digestive system and blood through the arteries. Cardiac muscle is heart muscle. Cardiac muscle is striated, or striped, like skeletal muscle but its contraction and relaxation are involuntary. The job of cardiac muscle is to pump blood.

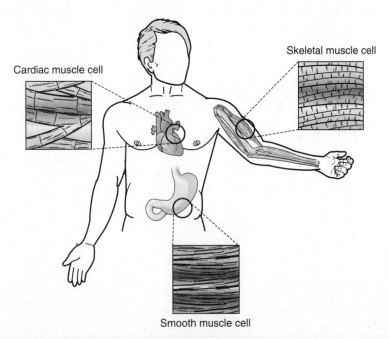

Cardiac muscle cell

Skeletal muscle cell

Smooth muscle cell

Problems of the muscular system include injuries, such as strains, in which muscles or tendons are stretched or torn. A muscle cramp is a painful and prolonged contraction of a muscle that may be triggered by nutritional deficiency, by an underlying medical condition, or which may be idiopathic (without a known cause). Many of the diseases that affect the muscular system affect other systems as well, particularly the nervous and skeletal systems. Muscular dystrophy is a group of several genetic disorders that cause weakness and wasting of different muscle groups of the body. The degree of disability and any reduction in life span associated with muscular dystrophy is related to the affected muscle groups.

Cardiovascular System

The cardiovascular system includes the heart and the blood vessels. The heart acts as a pump and the arteries, veins, and capillaries serve in the transport and delivery of blood, with all the oxygen and nutrients it carries, to all the cells throughout the body. The heart is a strong muscle that doesn't get tired or take a vacation or a nap. It beats approximately 72 times per minute in adult human beings, about 37,843,200 times in one year and 2,838,240,000 times in 75 years. The heart has four chambers, or compartments, two on the right and two on the left, with two atria on the top and two ventricles on the bottom. Blood flows through the heart from the right atrium, to the right ventricle, to the lungs (to pick up oxygen and get rid of carbon dioxide), then back in to the heart's left atrium, left ventricle, and out to the body cells. Arteries carry blood away from the heart and veins carry blood back toward the heart. Capillaries are the sites of oxygen and nutrient exchange into the cells and carbon dioxide and waste removal from the cells.

Heart disease is a leading cause of death in the United States for both men and women. There are many risk factors for heart disease including genetics, hardening of the arteries, obesity, high cholesterol and triglycerides, diabetes, lack of exercise, and cigarette smoking. Heart pain may be caused by a reduced blood supply to the heart muscle, a condition called angina pectoris. A heart attack is also called a myocardial infarction, or the death of heart muscle tissue. Strokes can also be a result of cardiovascular disease. Strokes happen when there is a problem with blood supply to the brain. This can happen as a result of a clot blocking blood flow through vessels in the brain or a disruption in blood flow because of bleeding from a ruptured blood vessel.

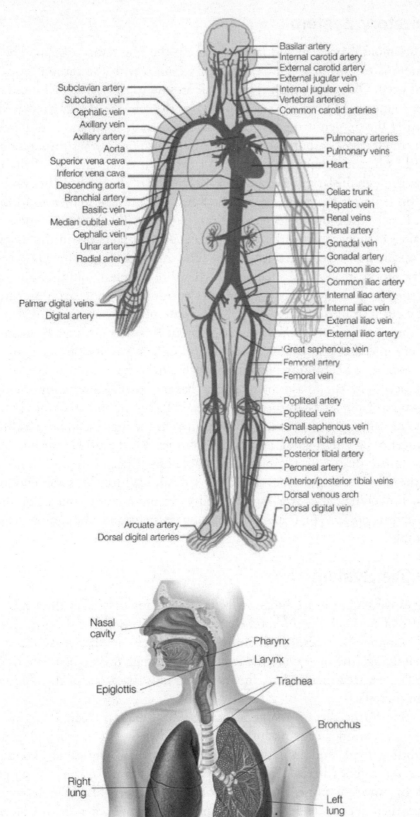

Basilar artery
Internal carotid artery
External carotid artery
External jugular vein
Internal jugular vein
Vertebral arteries
Common carotid arteries

Subclavian artery
Subclavian vein
Cephalic vein
Axillary vein
Axillary artery
Aorta
Superior vena cava
Inferior vena cava
Descending aorta
Branchial artery
Basilic vein
Median cubital vein
Cephalic vein
Ulnar artery
Radial artery

Pulmonary arteries
Pulmonary veins
Heart

Celiac trunk
Hepatic vein
Renal veins
Renal artery
Gonadal vein
Gonadal artery
Common iliac vein
Common iliac artery
Internal iliac artery
Internal iliac vein
External iliac vein
External iliac artery

Palmar digital veins
Digital artery

Great saphenous vein
Femoral artery
Femoral vein

Popliteal artery
Popliteal vein
Small saphenous vein
Anterior tibial artery
Posterior tibial artery
Peroneal artery
Anterior/posterior tibial veins
Dorsal venous arch
Dorsal digital vein

Arcuate artery
Dorsal digital arteries

Nasal cavity
Pharynx
Larynx
Trachea
Epiglottis
Bronchus
Right lung
Left lung
Diaphragm

Respiratory System

The nose, mouth, trachea, and lungs make up the respiratory system. The job of this system is to exchange the oxygen that we need with the carbon dioxide that we don't need. Oxygen is supplied by the air around us. All of our cells need oxygen in order to metabolize and use sugar; this gives the cells the energy they need to do work. After the oxygen is used, it bonds with a carbon atom and is disposed of as waste in the form of carbon dioxide. Air travels into our bodies through our nose or mouth; when it reaches the back of our throat, it enters the top of the trachea and goes through the larynx, or voice box. At the end of the trachea, the two bronchi branch off, entering the lungs. Inside the lungs, the bronchi branch to form bronchioles. Finally, the bronchioles end in alveoli, tiny sac-like structures where oxygen in the air is exchanged for carbon dioxide. All of the branching in the lungs looks like an upside-down tree; the trachea is like the trunk, the larger branches (bronchi) split off to form smaller and smaller branches (bronchioles), and then the smallest branches end in leaves (alveoli). Capillaries surrounding the alveoli exchange the needed oxygen from the air for the cellular waste product, carbon dioxide.

Common respiratory disorders include bronchitis and asthma. Bronchitis is a respiratory infection that causes irritation and inflammation of bronchi and bronchioles. Asthma is a condition in which bronchioles both constrict and secrete thick, excess mucus. The rate of asthma diagnoses has increased dramatically among children living in New Jersey in the last few years. Pollution and smoking, both first- and second-hand, have significantly contributed to the increase in both asthma and lung cancer cases. An ongoing respiratory disorder called COPD, chronic obstructive pulmonary disease, is a group of incurable lung diseases that make it difficult to breathe. The two major types of COPD are chronic bronchitis and emphysema. Chronic bronchitis causes swelling of the airways and excessive mucus production. Emphysema is a disease that causes loss of lung elasticity and a buildup of scar tissue in the lungs.

Digestive System

Our digestive system is a tubular collection of organs that takes the large macro-molecules of nutrients in the food that we eat and breaks them down into their smallest monomers. These small nutrient molecules can then pass through the intestinal lining, into the capillaries, to circulate through the bloodstream until they reach cells that need them. Here, the molecules cross back out of the blood vessels and into the cells.

The tube of the digestive system, called the **alimentary canal**, begins at the mouth, with its tongue for pushing food around, its teeth for slicing, crunching and grinding food, and its salivary glands for adding lubricants and digestive enzymes for carbohydrates. The mashed up food travels down the esophagus, pushed by wavelike muscular contractions, called **peristalsis**. Food enters the stomach where more muscular contractions churn it, and where hydrochloric acid and enzymes are secreted to begin protein digestion and to continue digestion of carbohydrates. The beginning of the small intestine receives more digestive enzymes to start breaking down lipids and it receives neutralizing fluids from two accessory organs, the liver and pancreas, to buffer the acidic stomach contents. The rest of the small intestine is for continued digestion and absorption of nutrients. By the time the food molecules enter the large intestine, they are basically a suspension of waste in water. The healthy large intestine absorbs back most of the water. Waste, called feces, is released from the body at the anus.

Common problems of the digestive system include diarrhea and vomiting. Diarrhea may seem to be simply an annoyance but it can lead to rapid dehydration and it can be an indicator of parasitic and bacterial infections. Vomiting is another sign of illness and is essentially reverse peristalsis. There are a number of inflammatory bowel diseases that cause the intestines to become inflamed (red and swollen). These chronic disorders, including Crohn's disease and ulcerative colitis, may cause pain, cramping, diarrhea, weight loss, and intestinal bleeding.

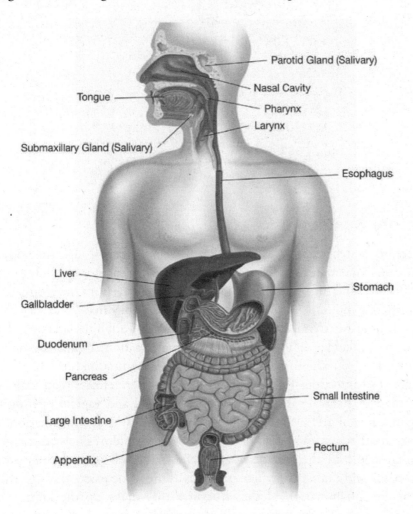

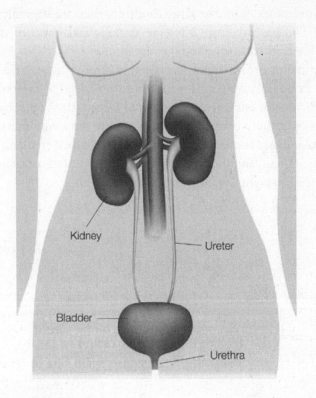

Excretory System

The excretory system eliminates waste that has circulated through the body via the blood stream. The main organs of the excretory system include the kidneys, ureters, urinary bladder, and the urethra. The kidneys filter toxins from the blood, collecting these poisonous molecules and adding water so that they may be carried safely out of the body as urine. Urine travels from the kidneys through the ureters to be stored in the urinary bladder. The process known as urination, or **micturition**, is the release of urine from the urinary bladder through the urethra.

Urinary tract infections (UTIs) are more common in females than males, mostly because there is a shorter distance from the urethral and vaginal openings to the anus. Symptoms of a UTI include cloudy urine and urgency of urination. Kidney stones are another common disorder of the urinary system. They can range in size from microscopic to the size of small driveway gravel. Most kidney stones are calcium-based and cause pain which increases as the stone passes through the ureter. Nephritis, an inflammation of the structures that make up the kidney, is often caused by an infection. Some signs of nephritis include lower back pain, scant urine, body swelling, and bloody urine.

Nervous System

The nervous system consists of the brain, the spinal cord, and the many nerves that extend into all parts of the body. Nerve cells, or **neurons**, throughout the body carry nerve impulses, from the brain to the muscles, from sensory organs and receptors to the spinal cord and brain, from the brain to our internal organs, and back again. Nerves provide us with sensory observations of pain, pressure, pleasure, and temperature. They transmit visual, **auditory** (hearing), **olfactory** (smell), and **gustatory** (taste) sensations. Inside the brain, nerves allow us to store information as memory, to speak and understand language, to coordinate our movements, to anticipate and to plan, and to think about concrete information as well as abstract ideas.

Problems of the nervous system include stroke, multiple sclerosis (MS), and Alzheimer's disease. Strokes usually happen when blood flow (and therefore oxygen) to the brain is somehow blocked, either by a clot or by a bleeding episode that diverts blood away from one part of the brain. When deprived of oxygen, brain cells die and will not be replaced, resulting in brain damage that may include reasoning, speech and memory deficits, paralysis, coma, and potentially death. MS is a chronic and unpredictable disorder of the central nervous system and MS begins with the formation of plaques, or inflammation in the white matter of the central nervous system (CNS). Plaque appearance is followed by destruction of myelin and the onset of symptoms. Myelin is an insulating material on neurons. When myelin disappears, the impulse traveling along a neuron may be distorted or lost completely. A diagnosis of multiple sclerosis indicates that some degree of neuromuscular communication is likely to be lost causing vision disturbances, abnormal sensations, compromised coordination, and difficulty walking. Alzheimer's disease is a common, degenerative neurological disease that causes dementia. Dementia involves the incurable deterioration and destruction of mental abilities. Alzheimer's patients exhibit confusion, and restlessness, as well as impaired memory, judgment, behavior, communication, and thought functions.

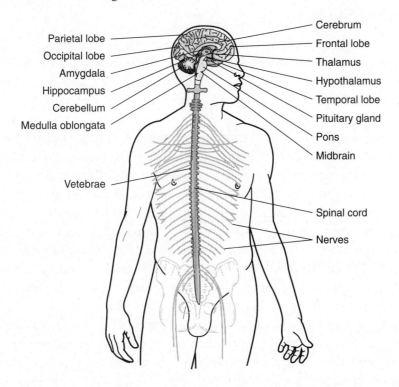

Major Endocrine Glands

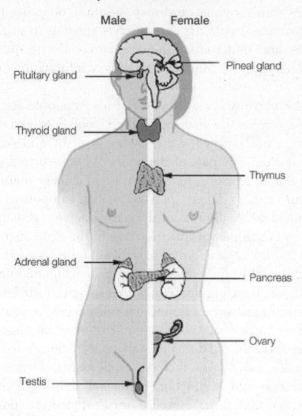

Male   Female

Pituitary gland

Pineal gland

Thyroid gland

Thymus

Adrenal gland

Pancreas

Ovary

Testis

Endocrine System

The endocrine system is the system of **hormones**, where they're made, how they're transported and what they affect. Hormones are chemicals that are secreted by one organ but that have their effect on a different organ. Some glands secrete hormones directly into the body fluids surrounding them; these are endocrine glands. Other glands secrete hormones into tubes, or ducts, to carry them to a specific destination; these are exocrine glands. One hormone, adrenaline, is produced by the adrenal glands which sit on top of the kidneys. Adrenaline is released by these glands when a person becomes frightened, or would benefit from a fight or flight response. For short amounts of time, adrenaline speeds the heart and breathing, increases blood pressure, stops digestion, dilates the pupils and, generally lets us run faster and jump higher than we ever could do without it. The pituitary gland is a master gland, telling other endocrine glands when to become active; the pituitary also controls growth. The thyroid gland influences the rate of the body's cellular reactions, parathyroid glands support calcium absorption, the pancreas produces insulin, and the ovaries and testes produce sex hormones, estrogen and testosterone.

Diabetes is a common endocrine disorder. There are two kinds of diabetes, Type I and Type II. Type I diabetes occurs when the pancreas does not produce insulin, a hormone needed to convert carbohydrates into the energy that our cells need to keep us alive. In order to metabolize and use sugars, people with Type I diabetes need insulin therapy. Type II diabetes is more common than Type I. A patient with Type II diabetes produces insulin but they either don't make enough of it or their cells don't acknowledge the insulin that is there. Symptoms of diabetes include excessive thirst, frequent urination, abnormal hunger, weight loss, irritability, and extreme fatigue. Endocrine disorders also include many reproductive system problems, hirsutism (excessive amounts of body hair), and pancreatitis (inflammation of the pancreas).

Reproductive System

The human reproductive system is designed to produce unique and fertile offspring. The male reproductive system continually produces sperm in the testes. It stores and transports them in the ducts of the epididymus, vas deferens, seminal vesicles, and urethra. The male reproductive system produces semen in which sperm is nourished, energized, and protected using the secretions of the seminal vesicles, prostate, and the bulbourethral glands. During fertilization, semen is transported through the urethra to be deposited as close as possible to a viable egg cell. The female reproductive system includes ovaries, where eggs (present since before the female was born) develop, and fallopian tubes for movement of the egg toward the uterus. The uterus is the organ where a fertilized egg may be implanted and will grow throughout a 40-week pregnancy. The uterus opens through the cervix and into the vagina, where sperm may be deposited during sexual intercourse. Female hormones, estrogen, progesterone, and follicle stimulating hormone (FSH) influence maturation and release of fertile eggs. These hormones also control the uterine environment to make implantation of a fertilized egg and fetal development possible. These hormones dictate whether the uterine lining will remain intact to accept and protect a fertilized egg, or if the lining will slough off during menstruation to provide a fresh environment the following month.

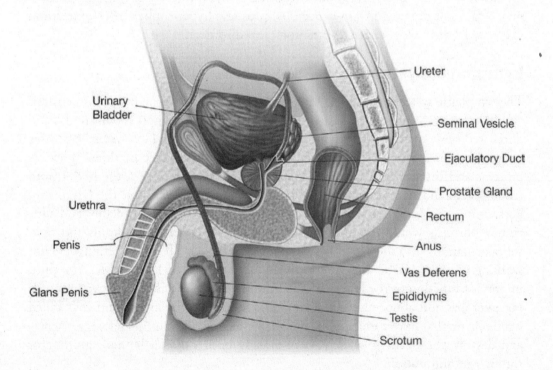

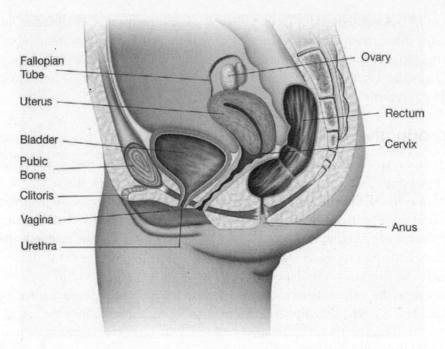

Problems with the reproductive systems include infertility. Infertility can occur in both males and females and may involve abnormal hormone secretion, sexually transmitted diseases, or the presence of diseased tissue. Endometriosis is a disorder of the female reproductive system that involves the growth of the type of cells that line the uterus (endometrial cells), on the outside of the uterus. Endometriosis can exist without obvious symptoms, or may cause pain, scarring, and fertility problems. In the male reproductive system, an enlarged prostate can cause difficulties with urination or it may be asymptomatic (causing no symptoms). An enlarged prostate may be a result of benign (non-cancerous) prostatic hypertrophy (BPH), infection, diabetes, heart disease, cancer, or neurological disorders.

Lymphatic System

The lymphatic system is an important part of your ability to maintain homeostasis and fight off pathogens. It is made up of lymphatic fluid, lymph vessels, nodes and ducts, red bone marrow, the thymus gland, and the spleen. This system has three major jobs. First, lymphatics maintain the body's fluid balance by draining excess proteins and **interstitial** fluid (loose fluid that bathes the cells of the body) from tissue spaces and it returns these products to the blood. Without the lymphatics performing this function, our body tissues would swell uncontrollably and ultimately our cells would die off. Secondly, lymphatic vessels move lipids and lipid soluble vitamins into the blood from the digestive system. If this function were not working we wouldn't be able to effectively use vitamins A, D, E, or K. The third important function of the lymphatics is to initiate very specific responses to invading microbes and abnormal cells. Lymphatic immunity involves blood cells called lymphocytes that have a memory of previous infection and can, therefore, recognize and destroy pathogens to which we've already been exposed, either naturally or through immunization.

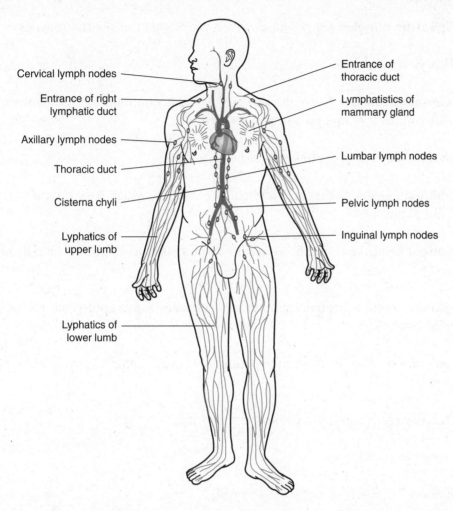

Cervical lymph nodes

Entrance of right
lymphatic duct

Axillary lymph nodes

Thoracic duct

Cisterna chyli

Lyphatics of
upper lumb

Lyphatics of
lower lumb

Entrance of
thoracic duct

Lymphatistics of
mammary gland

Lumbar lymph nodes

Pelvic lymph nodes

Inguinal lymph nodes

Autoimmune diseases are a problem because they indicate that a person's immune system identifies the persons own cells as foreign and attacks them. Auto-immune problems include allergic reactions and lupus. Allergies involve varying levels of overreaction to a substance to which other people do not react. Some common allergens include pollen, dust mites, molds, eggs, wheat, shellfish, tree nuts, medications, and poison ivy. Allergic reactions may range from a runny nose and red eyes, to constriction of the airways and excess mucus production in the lungs, to anaphylactic shock, which is a life-threatening emergency. Lupus is a chronic autoimmune disorder that affects multiple body systems. Symptoms of lupus include fatigue, joint pain, low fever, hair loss, weight loss, and enlarged lymph nodes. Additional problems with the lymphatic system are swelling and the inability to fight infection.

Important Terms to Remember

Alimentary canal: Tube of the digestive system; made up of the mouth, esophagus, stomach, small and large intestines.

Anatomists: Scientists who study how the body is put together.

Antagonistic muscles: Opposing members of each pair of skeletal muscles.

Auditory: Pertaining to the sense of hearing.

Autoimmune disease: A disorder in which a person's immune system identifies the person's own cells as foreign and attacks them.

Gustatory: Pertaining to the sense of taste.

Hormones: Chemicals that are secreted by one organ which have their effect on a different organ.

Interstitial fluid: Loose fluid, similar to blood plasma that bathes the cells of the body.

Keratin: A structural protein found in hair and nails, and in epidermal cells making them waterproof.

Ligament: Extension of the bone's periosteum; this tough connective tissue holds bone to bone.

Micturition: Urination; the elimination of urine.

Neuron: Nerve cell.

Olfactory: Pertaining to the sense of smell.

Peristalsis: Wavelike muscular contractions that move food in one direction through the alimentary canal.

Practice Questions

Multiple-Choice

1. You just ate a big lunch including a cheeseburger, fries, salad, and an ice cream sundae for dessert. Which of the following locations will NOT initiate any new digestive activity?

   (A) small intestine
   (B) stomach
   (C) mouth
   (D) large intestine

2. While you were eating, waves of muscular contractions helped move food into your stomach. These contractions are called _____.

   (A) peristalsis
   (B) metabolism
   (C) excretion
   (D) striation

3. Which of the following is not a job of your integument?

   (A) excretion
   (B) oxygen exchange
   (C) physical sensation
   (D) protection

4. Keratin is a structural protein in your hair and nails. What is its major function in the skin?

   (A) insulation
   (B) gives the skin its color
   (C) manufacture of vitamin D
   (D) waterproofing

5. What are the most important gases exchanged in the lungs? _____ entering the body and _____ leaving the body.

   (A) oxygen; nitrous oxide
   (B) carbon dioxide; carbon monoxide
   (C) oxygen; carbon dioxide
   (D) oxygen; carbon monoxide

6. There are two tubes in the throat. The one in the back is the _____ and it is used for _____.

   (A) small intestine; absorption of food
   (B) esophagus; air exchange
   (C) trachea; movement of food
   (D) esophagus; movement of food

7. In which of the following will your lymphatic system not be involved?

   (A) reducing tissue swelling
   (B) transport of vitamins A and D
   (C) destroying pathogens on first exposure
   (D) destroying pathogens following prior exposure

8. The heart is a unique muscle because it _____

   (A) contracts regularly, about 72 times per minute, as long as you're alive.
   (B) contracts regularly, about 20 times per minute, as long as you're alive.
   (C) doesn't interact with the skeleton.
   (D) is made of smooth muscle.

9. Blood circulates through the cavities of the heart in the following order:

   (A) left atrium, left ventricle, out to lungs, back to right atrium, right ventricle
   (B) left atrium, right atrium, out to lungs, back to right ventricle, left ventricle
   (C) right atrium, left atrium, out to lungs, back to left ventricle, right ventricle
   (D) right atrium, right ventricle, out to lungs, back to left atrium, left ventricle

10. Which of the following is not a muscle type?

   (A) skeletal
   (B) circular
   (C) cardiac
   (D) smooth

Free-Response

1. Why are the cardiovascular and respiratory systems considered to be very closely linked in function?

2. What impact does the endocrine system have on the rest of the body?

3. Why are some diseases and disorders included in several body systems?

Answer Explanations

Multiple-Choice

1. **D** The large intestine is the segment of the alimentary canal where water is reabsorbed and waste is consolidated. No new digestion happens in the large intestine. Carbohydrates begin digesting in the mouth, proteins begin digesting in the stomach, and lipids begin digesting in the small intestine.

2. **A** Peristalsis is the name for the waves of muscular contraction that move food through the alimentary canal. Metabolism is the sum of chemical reactions that occur in the bodies of living things. Excretion is the process of disposal of bodily waste. Striation is a pattern with stripes.

3. **B** Oxygen exchange is not a function of the skin. Functions of the integument do include excretion, physical sensation (touch, pressure, pain), and protection.

4. **D** Keratin, in the epidermis, provides the skin with waterproofing. Melanin gives the skin its color, manufacture of vitamin D is a complex chemical process that occurs in the skin but is not associated with keratin, and insulation is provided by subcutaneous fat underneath the dermis.

5. **C** Oxygen entering the body and carbon dioxide leaving the body are the two most important gases exchanged in the lungs.

6. **D** The posterior tube in the throat is the esophagus and its function is the movement of food from the mouth to the stomach.

7. **C** Lymphatics do not destroy pathogens upon first exposure, they destroy pathogens that they recognize from a previous exposure. Lymphatics are involved in reduction of tissue swelling and in the transport of vitamins A and D.

8. **A** The heart beats about 72 times per minute for most of your lifetime. Twenty beats a minute would not sustain life, smooth muscle doesn't interact with the skeleton either, and its made up of cardiac muscle.

9. **D** Blood circulates through the heart in the following sequence: Right atrium, right ventricle, left atrium, left ventricle.

10. **B** Circular muscle is not a type of muscle. Striated/skeletal, cardiac, and smooth are all types of muscle.

Free-Response Suggestions

1. **Why are the cardiovascular and respiratory systems considered to be very closely linked in function?**

   The cardiovascular and respiratory systems considered to be very closely linked in function because neither works without the other. A primary function of the blood is transport of oxygen to all the cells of the body and the removal of the waste product of cellular metabolism, carbon dioxide. The exchange of these two gases occurs in the alveoli of the lungs. The transportation of these two gases throughout the body occurs throughout the circulatory system. The exchange of these gases occurs at the capillaries.

2. **What impact does the endocrine system have on the rest of the body?**

   The endocrine system impacts the rest of the body because its glands secrete hormones that control, activate, or otherwise alter the actions of other organs. For example, human growth hormone secreted by the pituitary gland triggers skeletal growth, and adrenaline secreted by the adrenal glands causes increased respiratory rate, heart rate, and blood pressure.

3. **Why are some diseases and disorders included in several body systems?**

Some diseases and disorders are included in several body systems because they have symptoms that pertain to different systems, or because the disorder is influenced by multiple systems. For example, multiple sclerosis could be identified as disease of the muscular or nervous system, or as an autoimmune disease.

Ecology

Outline for Organizing Your Notes

I. Ecology
   A. Biotic Factors
   B. Abiotic Factors

II. Levels of Ecological Organization
   A. Organism
   B. Population
   C. Community
   D. Ecosystem
   E. Biome
   F. Biosphere

III. Living Arrangements
   A. Habitat
   B. Niche

IV. Feeding Arrangements
   A. Food Chain
   B. Food Web

V. Relationships Between Organisms
   A. Intraspecific
   B. Interspecific

VI. Energy Relationships
   A. Pyramid of Numbers
   B. Pyramid of Biomass
   C. Pyramid of Energy

VII. Succession
   A. Primary Succession
   B. Secondary Succession

VIII. Environmental Cycles
   A. Water Cycle
   B. Carbon Cycle
   C. Nitrogen Cycle

IX. Population Ecology
   A. Carrying Capacity
   B. Issues in Population Density

How are you like someone your age who lives in West Africa? How are you dissimilar to them? What factors influence these similarities and differences? You might say that many of the differences are cultural, but how much of culture is guided by our environment, by the living and nonliving components of our surroundings? How do these environmental factors influence each other? We humans are unique in that we have more control over our environment than any other species. So, how is the life of your pet cat or the tree growing in your yard different than the life of a cat or tree in another part of the world? The study of interactions between living organisms, and between these organisms and their surroundings is called **ecology**. Scientists who study the interactions between living things and their environments are called ecologists.

Biotic and Abiotic Factors

What are the things that influence living organisms? Well, they can all be broken down into two categories, living factors, called **biotic factors**, and nonliving factors, called **abiotic factors**. Biotic factors are all the elements of an organism's environment that are living, that once were living, or that are wastes or remains of living things. Biotic factors include species interactions (such as cooperation, parasitism, and predation), disease transmission, trees providing shelter and food for other organisms, and fungi increasing the nutrients in the soil through decomposition of organic material.

Abiotic factors are the physical and chemical elements of an organism's environment that have an impact on that living thing's existence. Some examples of abiotic factors include temperature, precipitation, water availability, soil quality, water quality and pH, soil type, rock types, light availability and intensity, and dissolved gases. In the course of any organism's efforts to stay alive, both biotic and abiotic factors have an impact.

Levels of Ecological Organization

Do you remember identifying the levels of organization of living things? We discussed how cells were the smallest independent units of living material and that cells, grouped together with other cells that had similar structures and functions, were called tissues. Similar tissues grouped together are organs, organs grouped together are organ systems and organ systems grouped together are organisms. Well, we can continue that organizational process to understand the relationships of organisms to each other and to the world in which they live. A group of organisms of the same species that live in an area together are called a population. Populations of different species living in an area are called a community. Interacting communities and the environments that they live in are called ecosystems. The varieties of ecosystems that comprise similar climatic regions on Earth are called biomes and all the biomes on the Earth are collectively called the biosphere.

Habitats Versus Niches

Habitats are living arrangements, places where organisms make their homes. One example of a habitat would be a forest of pine trees; here you are likely to find (in addition to the pine trees), mosses, mushrooms, worms, insects, slime molds, birds, chipmunks, squirrels, and deer. Another example of a habitat might be in the sand dunes at the shore. Here you might find insects and spiders, sand crabs, dune grass, mice, and birds. Habitat is simply an explanation of the surroundings in which you may find a particular population.

A **niche** is the description of how an organism fits into its environment. Explaining a niche is much more involved than explaining a habitat because it involves all the elements of the organism's existence as they relate to the other organisms that share that environment. So, for example, the niche of a rabbit that lives in the woods near your school would include many elements for your consideration. What is the rabbit's role in the food web? What does it eat and what eats it? Does its waste change the soil chemistry? Does its burrow prevent plants from growing, thus altering photosynthetic patterns, development of sheltering plants, and modifying food availability? Finally, what are the effects of many rabbits on this specific habitat, since rabbits tend to reproduce large quantities of offspring relatively quickly?

Several terms are available to help us identify the roles of different organisms in their niches. Organisms that conduct photosynthesis, like plants, green algae, and blue-green bacteria, are called producers. A producer is considered to be an **autotroph**, or an organism that makes its own food. A consumer is a **heterotroph**, an organism that gets its nutrients from eating another organism. A heterotroph may be an herbivore, exclusively eating producers, an omnivore eating both plants and animals, or a carnivore eating exclusively meat. Organisms that feed on decaying organic matter are called decomposers. Each of these organisms plays an important role in their own niche.

Food Chains and Webs

Food chains are linear; they simply tell you a sequence of organisms that use one another as a food source. An example of a food chain would be grass, which is eaten by a grasshopper, which is eaten by a bird, which is eaten by a snake, which is eaten by a hawk. In this food chain the grass is the producer because it conducts photosynthesis and produces 100 percent of the energy that can be produced in the form of carbohydrates. The grasshopper is an **herbivore**, it is the primary (represented by 1°) consumer in this food chain because it receives its food directly from the producer. The bird is an **omnivore** and is the secondary (2°) consumer in this food chain. The snake is a **carnivore** and is called the tertiary (3°) consumer. The hawk is also a carnivore; it is the quaternary (4°) consumer and is also called the **apex predator**. An apex predator is the ultimate carnivorous predator in any food chain. Apex predators, in other ecosystems, include lions, tigers, and great white sharks.

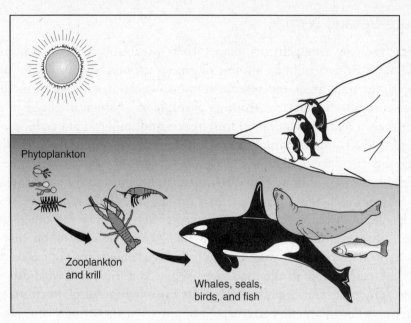

A Polar Food Chain

Food webs are more complex than food chains. They include several or many different food chains in an effort to show all the possible feeding relationships that exist in an ecosystem or all of the possible feeding connections for specific organisms. In the grassland food chain example, other species that might be found eating the grass include rabbits, deer, squirrels, and mice. Other levels of consumers would then be modified in a different feeding relationship; a tertiary consumer might become a secondary consumer, one carnivore might feed on several types of omnivore, or other apex predators might emerge. Another role in a food web is that of a **keystone species**, a species that is critically important to its ecosystem. In fact, a keystone species is disproportionately more important to its food web than its population density suggests. If a keystone species disappears from its environment, other species in the ecosystem are likely to die off. An example of a keystone species would include elephants in the African savannah. The grasslands in Africa stay grassy primarily because the activity and movement of elephants prevents shrubs and trees from growing. This prevents woody trees from growing over the grassland ecosystem. If the elephants weren't there, the grasses would also disappear and all the grassland inhabitants would either die off or move away.

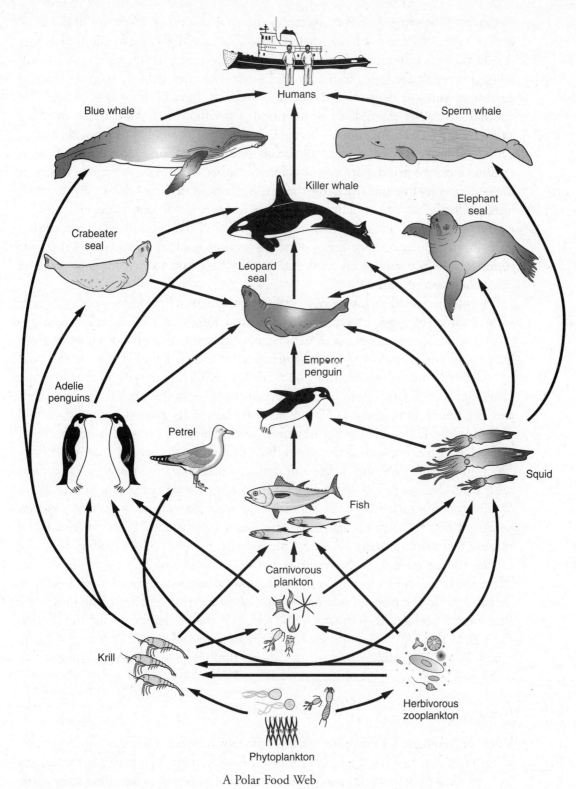

Blue whale

Sperm whale

Killer whale

Crabeater seal

Elephant seal

Leopard seal

Adelie penguins

Emperor penguin

Petrel

Squid

Fish

Carnivorous plankton

Krill

Herbivorous zooplankton

Phytoplankton

A Polar Food Web

Species Interactions

One of the biotic factors that requires extra attention would be species interactions. There are several types of species interactions that can be divided into two groups: **intraspecific** and **interspecific** relationships. Intraspecific relationships exist when members of the same species compete for the same resources. Members of the same species may compete for food, shelter, space, a mate, light, or for dominance. A simple example of competition might involve two trees growing right next to each other competing for light, nutrients, and water. One tree is likely to grow taller or develop a wider or deeper root system in order to have better access to available resources. Another example of competition is the fighting behavior seen in many animal species when individuals battle each other for the rights to food, shelter, or a mate. Another type of intraspecific behavior is cooperation. An example of cooperation might be activity in an ant colony or a bee hive where many members of the same species live and work together for the benefit of their colony. One extreme intraspecific behavior is infanticide by male lions. A newly dominant male lion in a pride (a group of lions that lives together) will kill the cubs already born to that pride in order to encourage the lionesses to become sexually receptive to him. This eliminates the progeny of his predecessor and allows him to pass his genes on to a new generation of lion cubs.

Interspecific relationships exist between members of different species. There are a number of types of symbiotic (close and long-term) interspecific relationships, including mutualism, commensalism, parasitism, and predation. Mutualism happens in a relationship when both members of the relationship benefit from the interaction. A mutualistic relationship exists between a honey bee and the flowering plant that it pollinates. Both of these organisms benefit; the bee gets nourishment and the flower is pollinated. Commensalism occurs in a relationship when one organism benefits and the other organism doesn't benefit, but it also isn't harmed. The relationship between a tree and a bird nesting in the tree is commensalistic. The bird benefits from the shelter and structure offered by the tree, it may even eat seeds from the tree but most nesting birds do not provide either a benefit or harm to the tree in which they live. Parasitism is a relationship in which one species benefits while the other is harmed. Rarely is the parasitized (host) species killed because that would usually also harm the parasitic member of the relationship. The human disease malaria is the result of a parasitic relationship between the protist *Plasmodium* sp. and its human host. This malarial parasite is transmitted from one human host to the next by way of a vector, or an intermediate host, the mosquito. Predation is a relationship in which one species is the predator (seeking to kill food for itself), while the other species is its prey (a potential source of food). When a grizzly bear eats a salmon that it just caught, the bear is the predator and the fish is the prey.

Energy Flow

When an ecologist is examining the relationships between species in an ecosystem, it becomes obvious that there are proportionate relationships between the role of the organism in its environment, how many members of the species exist, and how much energy they exchange. An easy way to identify those proportionate relationships is to diagram the populations in terms of a pyramid. Ecologists use three main pyramid models: the **Pyramid of Numbers**, the **Pyramid of Biomass**, and the **Pyramid of Energy**. The Pyramid of Numbers tells us that as we move up from producers to primary, to secondary, and tertiary consumers, there are significantly fewer members of the populations that make up each higher level. The Pyramid of Biomass indicates that as we move up these same levels, the biological mass of the organisms is reduced by approximately 90 percent at each level. The Pyramid of Energy indicates that as we travel up each level, the energy available to the consumers is reduced by about 90 percent.

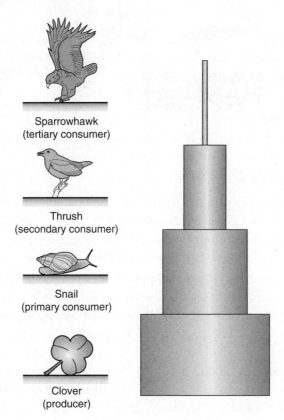

Pyramid of Numbers

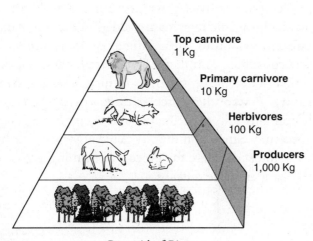

Pyramid of Biomass

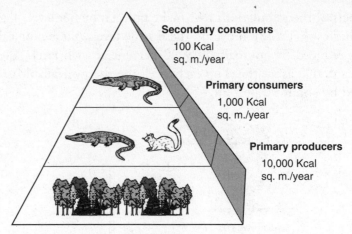

Secondary consumers
100 Kcal
sq. m./year

Primary consumers
1,000 Kcal
sq. m./year

Primary producers
10,000 Kcal
sq. m./year

Pyramid of Energy

Communities and Succession

Another element of ecology is the understanding of how communities and ecosystems develop over time. The development of a healthy ecosystem is a result of the stability of the land form combined with the interactions of the organisms and communities that live there. Changes that happen in an ecosystem over time are called ecological succession. Abiotic changes may happen as a result of natural events like a fire, flood, earthquake, volcanic eruption, or landslide or they may be a result of human activity like urban expansion, clear cutting forests, unethical waste disposal, or building dams. The populations living in an area are modified by all of these types of events.

Primary succession occurs in an environment where no soil is initially present. This may be following a volcanic explosion or formation of a new island. Without soil there will be no producers and, therefore, no consumers. Over time, grasses, moss, and lichens will embed in cracks in the rock and their root systems will begin the process of breaking up the surface rock and forming soil; these first organisms to grow here are called pioneer species. As more soil forms, larger plants will move in and as these die, their organic remains are added to the soil making it richer and attracting more diverse organisms to live there. As more producers grow in this environment, more consumers join them. Over decades and centuries, more diversity develops and finally, a stable ecosystem is formed. This ecosystem is considered a **climax community**. A **secondary succession** happens when an existing ecosystem, that already contains soil, is decimated and then goes through a progression of successions to become a climax community. A secondary succession often happens following a fire or a mudslide. Here the process of forming soil isn't an issue so the successions usually happen more quickly.

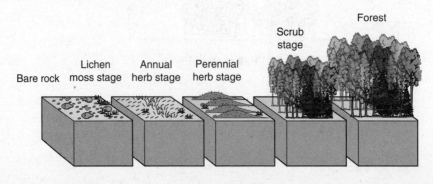

Forest

Scrub
stage

Lichen   Annual   Perennial
Bare rock   moss stage   herb stage   herb stage

Environmental Cycles

We have already examined the roles of several atoms and molecules that are crucial to the existence of life on Earth. It's important to review them now in the context of ecological interactions. Approximately 71 percent of the Earth's surface is covered with water so it makes sense that we understand the role of the Water Cycle in terms of its ecological impact. Since about 70 percent of the Earth's water is in the oceans, it would make sense to begin looking at the Water Cycle here. When sunlight hits the surface of the oceans, it warms the water causing evaporation of water vapor into the atmosphere. Water also enters the environment as a result of transpiration (the release of water through the leaves of a tree) from plants, cellular respiration, and excretion. Atmospheric water vapor condenses to form clouds and ultimately, returns to the Earth in the form of precipitation that may be stored on Earth as ice or as fresh or salt water. Life forms on Earth use these molecules of water temporarily and they are continually returned and exchanged through the Water Cycle.

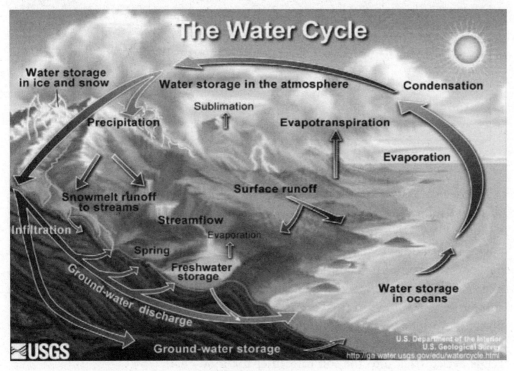

The Water Cycle

Carbon is constantly being cycled through the environment by way of the Carbon Cycle. Abiotic sources of carbon include atmospheric carbon dioxide, rocks, soil, and the burning of fossil fuels. Carbon is used by plants for photosynthesis and is released by consumers as a by-product of cellular respiration. Carbon is moved through the food chain as organisms eat and metabolize their food. Carbon is cycled through the soil when organisms die or leave waste and it decomposes into the ground.

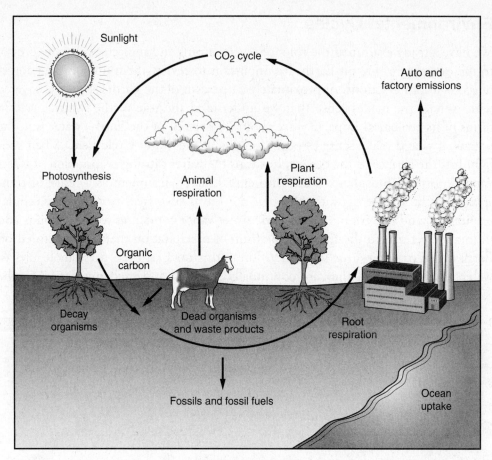

The Carbon Cycle

Nitrogen is another critical element in the bodies of all living things. Remember, all living things are made of proteins and proteins have amino groups that are nitrogen based. When any living thing dies, its molecules return to the environment. As the bodies of dead organisms (bacteria, protists, fungi, plants, and animals) break down, their nitrogen molecules are moved through the decomposition process in a specific sequence of events. Decomposers in the soil convert amino acids into ammonia, which is converted to ammonium, then nitrites, and finally, nitrates. Nitrates can then be taken up by plants to help them form proteins or they can be de-nitrified and returned to the atmosphere as simple nitrogen. Atmospheric nitrogen is cycled back into the soil to reenter the food chain.

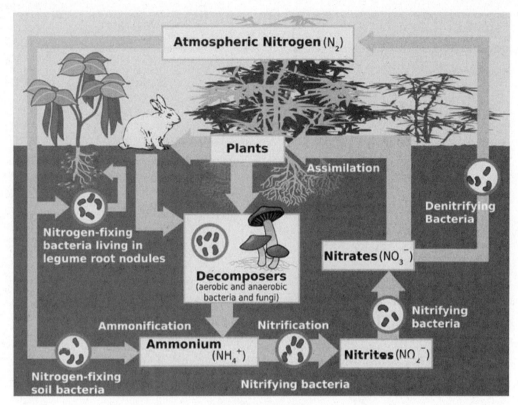

The Nitrogen Cycle

Population Ecology

Another element of understanding ecology is the study of the impacts and interactions between populations of organisms and with their environment. Remember that a population is a group of organisms of the same species living in the same environment. Population ecology investigates statistics on spatial distributions, population density, age, sex, birth rates, and mortality rates. A population ecologist will also be interested in what causes changes in any of these areas. In any environment, there will be a limit to the ability of that location to provide resources to its inhabitants. That limitation may be space, food, or water. It may be that as crowding intensifies, disease processes or extreme aggressiveness become commonplace. These conditions are called **limiting factors** and are problems that make it difficult or impossible for a population to live and grow in its environment.

Limiting factors are closely linked to **carrying capacity**, the number of organisms that an environment can sustain indefinitely. As a population density increases toward carrying capacity, birth rates often decrease and death, or mortality, rates increase. At the same time the population may see an increase in **emigration**, or movement out of the environment, and a decrease in **immigration**, population movement into an environment. Population ecologists often present these statistics in the forms of graphs. The development of a population in an environment that hasn't yet reached its carrying capacity will have the shape of the letter "J" and is called a **J curve**. When the population reaches the environment's carrying capacity, and the population density levels off, it takes on the shape of an **S curve**.

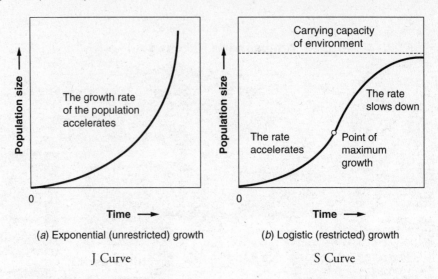

(a) Exponential (unrestricted) growth

J Curve

(b) Logistic (restricted) growth

S Curve

Important Terms to Remember

Abiotic factor: The non-living, physical, and chemical elements of an organism's environment that have an impact on that living thing's existence.

Apex predator: The ultimate carnivorous predator in any food chain.

Biotic factor: All the elements of an organism's environment that are living, that once were living, or that are wastes or remains of living things.

Carnivore: A consumer that eats primarily, or exclusively, meat.

Carrying capacity: The number of organisms that an environment can sustain indefinitely.

Climax community: A stable ecosystem formed over many years with large and diverse populations.

Ecology: The study of interactions between living organisms and between these organisms and their surroundings.

Emigration: Movement out of the environment by members of a population.

Habitat: An organism's living arrangements; the places where organisms make their homes.

Herbivore: A plant eater; a consumer that feeds directly on producers.

Immigration: Movement into an environment by members of a population.

Interspecific: Relationships within a community in which members of different species compete for the same resources; examples include predation, parasitism, commensalism, mutualism.

Intraspecific: Relationships within a community in which members of the same species compete for the same resources; examples include competition and cooperation.

J curve: The development of a population in an environment that hasn't yet reached its carrying capacity.

Keystone species: A species that is critically important to its ecosystem. If a keystone species disappears from its environment, other species in the ecosystem are likely to die off.

Limiting factor: Issues that make it difficult or impossible for a population to live and grow in its environment; examples include space, food, water availability, and disease.

Niche: How an organism fits into its environment; its feeding roles, its shelter requirements, and its interactions with other living things.

Omnivore: Consumers that eat both plant (producer) and animal (consumer) products.

Primary succession: An ecosystem where no soil is initially present; these tend to develop slowly because soil must be formed to allow extensive growth of producers.

Pyramid of Biomass: A model that tells us that as we move up from producers to primary, to secondary, and tertiary consumers, each level has predictably less organic matter.

Pyramid of Energy: A model which tells us that as we move up from producers to each level of consumer, 90 percent of the available energy is lost; only 10 percent of the available energy is passed from one level to the next.

Pyramid of Numbers: A model that tells us that as we move up from producers to primary to secondary and tertiary consumers, there are significantly fewer members of the populations that make up each higher level.

S curve: Graph that represents the leveling off of population density when a population reaches its environment's carrying capacity.

Secondary succession: When an existing ecosystem, that already contains soil, is decimated and then goes through a progression of successions to become a climax community.

Practice Questions

Multiple-Choice

1. You and your friends used to go swimming at a local pond during the summers but lately the pond is so dirty-looking and green. You bring a water sample from the pond back to your biology class and you and your teacher discover a large population of algae in the sample. Your teacher says it may be the result of an algae bloom. Which of the following terms best describes the algae in your water sample?

   (A) herbivore
   (B) consumer
   (C) biotic factor
   (D) abiotic factor

2. When you went to the pond to get your water sample you noticed a few turtles, frogs, birds, and lily pads at the pond. These different groups of animals living in the pond ecosystem are collectively called a(n) _____.

   (A) biome
   (B) ecosystem
   (C) population
   (D) community

3. Over the last few weeks, you and your friends have been going back frequently to explore the pond. Each time you see what seem to be the same three turtles sitting on the same logs and rocks. You are quite certain that the turtles make their homes right near their logs and rocks. This area would be considered the turtles' _____.

   (A) habitat
   (B) niche
   (C) food chain
   (D) Pyramid of Biomass

4. The algae and lily pads in the pond are photosynthetic organisms and are therefore _____, while the turtles and frogs get their nutrients from eating other organisms and as a result are called _____.

   (A) consumers; producers
   (B) herbivores; omnivores
   (C) autotrophs; heterotrophs
   (D) abiotic factors; biotic factors

5. In the following food chain, which organism is the primary consumer?

> clover → caterpillar → songbird → snake → hawk

(A) clover
(B) caterpillar
(C) songbird
(D) hawk

6. In this food chain, the hawk is the _____.

> clover → caterpillar → songbird → snake → hawk

(A) lead producer
(B) major omnivore
(C) major herbivore
(D) apex predator

7. Which of the following is an example of an intraspecific relationship?

(A) a human mother, father, and their baby
(B) an oak tree blocking the sunlight that reaches a young maple tree
(C) a female dog treating a kitten as though it were its own puppy
(D) a bee pollinating a rose

8. The relationship between a robin and the tree in which it builds its nest is _____.

(A) commensalistic
(B) mutualistic
(C) parasitic
(D) predatory

9. Which model shows us that as we travel up each feeding level, the chemical energy available to those consumers is only about 10 percent of that available to the lower level?

(A) Pyramid of Biomass
(B) Pyramid of Reductions
(C) Pyramid of Energy
(D) Pyramid of Numbers

10. On your hike through a Costa Rican rain forest, you see many tall trees that create a canopy over the forest. There is not a lot of undergrowth below them and you find it much easier to walk though this forest than through the woods back home in New Jersey. As the day goes by, you realize you've also not seen much sunlight. You realize that sunlight is a(n) _____ factor in the rainforest ecosystem.

   (A) unnecessary
   (B) interfering
   (C) limiting
   (D) biotic

Free-Response

1. Choose an ecosystem and describe a food web that is likely to exist there. Explain the feeding roles and niches for each organism in that ecosystem.

2. Explain the three model pyramids and describe how they relate to each other.

3. Explain three element cycles and how they impact ecosystems.

Answer Explanations

Multiple-Choice

1. **C** All the living things that have an impact on their ecosystem are called biotic factors. Algae is living and, therefore, not an abiotic factor. Algae is an autotroph and a producer and is, therefore, not a consumer or herbivore.

2. **D** A group of different species living together in an area is a community. A population is a single species within a community. An ecosystem contains both the communities and nonliving components in an area. A biome is all the ecosystems in a geographic region.

3. **A** The place where an organism lives is its habitat. The turtle's niche would be its role in its environment and its food chain would be the organisms that it eats and those that eat turtles. The turtle does not have its own Pyramid of Biomass; its ecosystem would be described by a pyramid model.

4. **C** The plants in the pond are autotrophs and producers and the animals in the pond are heterotrophs and consumers. Abiotic factors are nonliving.

5. **B** In this food chain the caterpillar eats the producer (clover), so the caterpillar is the primary consumer. The songbird is the secondary consumer. The hawk is the quaternary consumer.

6. **D** The hawk is the apex predator. The hawk is not a producer, omnivore, or herbivore.

7. **A** A human mother, father, and their baby exemplify a relationship between members of the same species (an intraspecific relationship). An oak tree and maple tree are two different species, as are a dog and a cat, and a bee and a rose. Two different species have interspecific relationships.

8. **A** The relationship between a robin and the tree in which it builds its nest is commensalistic because the robin benefits and the tree neither benefits nor is harmed. If they were mutualistic, both would benefit. If they were parasitic, one would benefit and one would be harmed. If they were predatory, one would be trying to kill and eat the other.

9. **C** The Pyramid of Energy is the model which shows us that as we travel up each feeding level, the chemical energy available to those consumers is only about 10 percent of that available to the lower level. The Pyramid of Biomass refers to the biological material in an ecosystem and the Pyramid of Numbers is about the number of members of different feeding levels in an ecosystem. There is no Pyramid of Reductions in ecology.

10. **C** Sunlight is a limiting factor in the rainforest ecosystem. Sunlight is necessary and is abiotic. Sunlight does not interfere with the rainforest ecosystem.

Free-Response Suggestions

1. **Choose an ecosystem and describe a food web that is likely to exist there. Explain the feeding roles and niches for each organism in that ecosystem.**

   A rainforest ecosystem would include many autotrophs/producers and would include a variety of canopy and understory trees, vines, shrubs, and flowering plants. Primary consumers/herbivores would include insects, parrots, monkeys, and small mammals, including bats. Secondary consumers and omnivores include amphibians, reptiles, insects, bats, and birds. Tertiary consumers include large snakes and jaguars. Some consumers, like the bats, fit into several feeding levels depending their food source. Habitats for all these organisms could be at any level of the rainforest; emergent layer, canopy, understory, shrub, and ground layers.

2. **Explain the three model pyramids and describe how they relate to each other.**

   The Pyramid of Numbers tells us that as we move up from producer, to primary, to secondary, and tertiary consumers, there are significantly fewer members of the populations that make up each higher level. The Pyramid of Biomass indicates that as we move up these same levels, the biological mass of all the organisms is reduced by approximately 90 percent at each level. The Pyramid of

Energy indicates that as we travel up each level, the energy available to the consumers is reduced by about 90 percent. The three Pyramids relate to each other because they all show how, as we move away from the producers, organisms get larger but fewer. Food becomes less energizing so more food must be consumed and more time spent catching it.

3. **Explain three element cycles and how they impact ecosystems.**

During the Water Cycle, sunlight hits the surface of the oceans and it warms the water, causing evaporation of water vapor into the atmosphere. Water also enters the environment as a result of transpiration from plants, cellular respiration, and excretion. Atmospheric water vapor condenses to form clouds and ultimately, returns to the Earth in the form of precipitation which may be stored on Earth as ice, or as fresh or salt water. Life forms on Earth use these molecules of water temporarily and they are continually returned and exchanged through the ongoing Water Cycle.

Carbon is used by plants for photosynthesis and is released by consumers as a by-product of cellular respiration. During the Carbon Cycle, carbon is moved through the food chain as organisms eat and metabolize their food. Carbon is cycled through the soil when organisms die or leave waste and it decomposes into the ground. When any living thing dies, its molecules (including carbon, oxygen, hydrogen, and nitrogen) return to the environment. As the bodies of dead organisms break down, their nitrogen molecules are moved through the decomposition process in a specific sequence of events. Decomposers in the soil convert amino acids into ammonia, which is converted to ammonium, then nitrites and finally, nitrates. Nitrates can then be taken up by plants to help them form proteins or they can be de-nitrified and returned to the atmosphere as simple nitrogen. Atmospheric nitrogen is cycled back into the soil to reenter the food chain.

CHAPTER 15

Biomes

Outline for Organizing Your Notes

I. Characteristics of Biomes

II. Water Biomes
    A. Oceans
    B. Coral Reefs
    C. Estuaries
    D. Wetlands
    E. Freshwater

III. Terrestrial Biomes
    A. Tundra
    B. Taiga
    C. Temperate Forest
    D. Rainforest
    E. Grasslands
    F. Desert

How would you describe the environment where you live? What kinds of populations live there? What might you say about the climate there? Have you ever thought about the influence of weather and geography on your community and on your own niche? What is your niche, anyway? Have you ever visited or lived in places that had significantly different weather or landforms? How might climate and landscape impact the organisms that live there? What might happen to the populations that live in a region if the weather or landforms were to change?

As we've already discussed, both biotic and abiotic factors impact the populations that live in any ecosystem. When we look at ecosystems around the world, we can identify patterns of similar biotic and abiotic factors based on the area where they're found and also on that area's specific climate and **topographical**, or geographic landscape, characteristics. All of these features, collectively, create a **biome**. A biome is a large geographical region that contains distinctive landforms and weather patterns; it also contains communities of interdependent and unique species that thrive under, and have adapted to, the conditions of that biome.

Criteria for Distinguishing Biomes

Biomes are groups of ecosystems that exist under the environmental conditions that are present in different parts of the world. The weather patterns and landforms in a biome influence the types of vegetation that grow there. The types and availability of vegetation will, in turn, impact the animal populations that will be found there. Living in any biome, in addition to plant and animal species, will be appropriately adapted populations of fungi, protists, and bacteria. All of the populations in a biome, regardless of size, complexity, or kingdom, are interdependent and will be impacted if an environmental change occurs or if any species is removed from that ecosystem.

Water Biomes

The Earth is estimated to be covered by more than 71 percent of its surface area with water and the remaining 29 percent of the earth's surface is considered to be terrestrial. Water biomes are first classified by whether they contain salt or fresh water. Salt water biomes make up the vast majority of the Earth's water biomes. Salt water biomes include oceans, coral reefs, and estuaries. Salt water biomes are inhabited by incredibly diverse populations including many species of marine algae that collectively produce a large percentage of the Earth's oxygen supply and that take in huge amounts of carbon dioxide from the atmosphere.

MARINE BIOMES—OCEANS

Oceans make up the largest biome on Earth. They contribute to the water cycle as their surface water evaporates, entering the atmosphere and then returning to Earth's surface in the form of precipitation. The world's oceans contain a salt concentration, or salinity, of approximately 3.5 percent. Oceans are divided into four zones: the **intertidal**, **pelagic**, **abyssal**, and **benthic** zones, which each contain unique and diverse ecosystems and species.

The intertidal zone is the area closest to shore, where ocean and land meet. This zone is strongly affected by waves and tides; it can have a gradual, smooth, and sandy transition from ocean to land or it can be treacherously rocky and mountainous. The intertidal zone is turbulent and always changing, so it's difficult for vegetation to establish itself. Temperatures can vary widely in this zone because the water is shallow enough to be affected by air temperature. The vegetation found in this zone primarily includes forms of algae, including green algae that grows on rocks and brown algae (seaweed). Animals that populate the intertidal zone include worms, crustaceans, clams, and shore birds.

The pelagic zone is the open ocean. Temperatures vary widely here because ocean currents are constantly mixing water of different temperatures. Producers found in this zone include surface seaweed and phytoplankton. Huge numbers of microscopic, floating organisms called **plankton** are found in the ocean; some are phytoplankton that conduct photosynthesis and mostly include members of the plantlike protist phylum. Animal species that live in the pelagic zone must either swim or float and include zooplankton, fish, dolphins, and whales. Zooplankton are primarily tiny arthropods (they are related to insects and lobsters) that float in the ocean currents. Krill, a tiny shrimp-like zooplankton, is one of the oceans major food sources.

The abyssal zone is the deepest part of the ocean, where sunlight doesn't reach. This zone is very cold, it is under high pressure, and it has sparse nutrient availability. Producers in this region are not photosynthetic, since they don't receive sunlight. Instead, they are chemosynthetic, using chemical energy, often from thermal vents. Most of the producers in this region are chemosynthetic bacteria that use hydrogen sulfide or methane to produce energy for themselves and the rest of their food web. Consumers that live in the abyssal zone include worms, shrimp, crabs, and fish.

The benthic zone is the sandy floor of the pelagic zone. Temperature and pressure in this region get colder as the water depth increases. The main vegetation in this area is seaweed. Animals in the benthic zone find this region to be nutrient-rich because of all the **detritus**, or organic waste, that settles to the ocean floor. This zone is inhabited by bacteria, sea stars, sea anemones, crabs, and fish.

MARINE BIOMES—CORAL REEFS

Coral reefs are mostly found in temperate or tropical warm water off the coasts of continents and islands throughout the world. The abiotic structure of coral reefs is composed of the calcium carbonate skeletal remains left behind by dead coral polyps. The biotic elements of coral reef structure are coral polyps, a simple class of animals related to jellyfish, and sea anemones, along with a photosynthetic form of algae. Other species of animals found in coral reefs include worms, crustaceans, sea stars, octopus, and fish.

MARINE BIOMES—ESTUARIES

An estuary is where rivers and streams join the ocean. This is a region where salt and fresh water come together, usually with a lot of mixing and tidal activity, resulting in very wide range of **salinity**, or salt concentration. Estuaries have a strong nutrient base including seaweed and other forms of algae as well as marsh grasses. Animals found in estuarine environments include worms, clams, oysters, crabs, and shore birds. Many young fish can be found in well-mixed estuaries; they have plenty of food and are somewhat protected by the seaweed and marsh grasses from the adult species of the pelagic zone that would prey on them in a less protected environment.

MARINE OR FRESHWATER BIOMES—WETLANDS

Wetlands are ecosystems that are covered in standing water for, at least, some of the year. Wetlands may be comprised of fresh or salt water. Marshes are wetland ecosystems that contain primarily soft, herbaceous plant life, while swamps contain more wood-based plant life. Wetlands contain very diverse species of plants and animals. Producers include grasses, cattails, sedges, cypress, and mangrove trees. Consumers receive nutrients from producers as well as from detritus. Animal species are diverse and include insects, worms, frogs, toads, snakes, alligators, birds, fish, rats, raccoons, opossums, and beavers.

FRESHWATER BIOMES

Water biomes that have salt concentrations of one percent or less are considered freshwater. Freshwater biomes include rivers, streams, lakes, and ponds. These biomes can have very different characteristics based on water movement, oxygen availability, depth, and temperature. Vegetation, or **flora**, varies based on water quality and depth; phytoplankton and zooplankton will be found at the surface while algae and aquatic plants are likely to be found in many freshwater biomes. **Fauna**, or animal species, in freshwater ecosystems include insects, worms, crustaceans, amphibians, birds, and fish.

Terrestrial Biomes

Land-based biomes are, like water-based biomes, dependent on temperature, nutrient availability, and landforms. Unlike conditions in marine and freshwater biomes, the availability of water, seasonal changes, and weather patterns play an important role in terrestrial biomes. Major terrestrial biomes include tundra, taiga, temperate forests, rainforests, grasslands, and deserts.

TUNDRA

The tundra is a northern hemisphere biome characterized by its lack of trees and its continually cold temperatures (averaging −18°F). Temperatures during the long winters can go as low as −94°F and during the summer can go as high as 60°F. The tundra has little precipitation (six to ten inches per year) usually in the form of snow, a short growing season (less than two months), minimal biodiversity, and poor nutrient availability. Tundra is the biome of the land masses surrounding the Arctic. A layer of **permafrost**, permanently frozen soil, exists year round, about one to three feet below the ground's surface. Trees and other large plants do not grow in the tundra because permafrost blocks them from developing a deep root system.

Plants that live in the tundra are primarily grasses, mosses, lichens, sedges, and shrubs. Ice melt during the short summers can cause the development of marshy tundra where thousands of insects (flies, mosquitoes, gnats, grasshoppers) grow and develop, becoming an important food source for resident and migrating birds (including snowy owls, gulls, arctic loons, and snow geese). Fish species in the tundra include salmon, trout, and cod. Mammals that live in the tundra include lemmings, arctic hares, squirrels, ermine, porcupines, caribou, wolves, and polar bears.

TAIGA

The taiga is another cold, Northern Hemisphere biome. It's found just to the south of tundra throughout the Eurasian and North American continents and is the largest terrestrial biome on Earth. Taiga is also called the "boreal forest"; it contains evergreen (cone- and needle-bearing) trees as its primary vegetation. Pine needles are leaves that are adapted to a cold environment; they don't have a lot of surface area to lose water and they're very well protected by a thick, waxy coating called a cuticle. Temperatures in the taiga range during the winter from −65° to 30°F and during the summer can go as high as 70°F. Winter lasts for a full six months in the taiga, while the summer lasts between two and three months. Total precipitation averages between one and three feet and is usually in the form of snow.

Flora of the taiga includes predominantly pine, spruce, hemlock, and fir, with some mosses and lichens on the forest floor. Much of the taiga is shaded, dense forest with nutrient-poor, acidic soil, which makes undergrowth development difficult. Fauna found in the taiga include worms, insects, salmon, trout, squirrels, weasels, deer, muskrats, moose, musk ox, foxes, badgers, beavers, and bears. Life is difficult in this cold biome but many migratory animal species make their temporary homes in the taiga during the warmer months of the year.

TEMPERATE FOREST

Deciduous forests can be found in both the Northern and Southern hemispheres. They are characterized by their four distinct seasons; winter, spring, summer, and fall. They are named after the trees that are their primary producers, deciduous trees, which shed their broad leaves during the fall of each year. Temperate deciduous forests range in temperature throughout the year (and according to altitude) from −22° to 82°F, with an average annual temperature of 50°F. Precipitation in the deciduous forest, in the form of rain or snow, averages 30 to 60 inches per year. Sunlight penetrates the temperate forest **canopy** allowing vegetation to grow below, helping to create a very nutrient-rich environment. Soil in the deciduous forest also tends to be rich with the decomposing remains of fallen trees and leaves, dead organisms, and other organic waste.

Deciduous trees that form a moderately dense canopy for this biome include maple, oak, elm, chestnut, and hickory. Other plants, found on lower levels (or **strata**) of forest height, include shrubs like mountain laurel, rhododendron, and azalea along with even lower strata-like mosses and lichens. Animal life is diverse in temperate forests and includes worms, insects, frogs, birds, mice, rabbits, squirrels, raccoons, opossum, foxes, deer, and bears.

RAINFOREST

Rainforest biomes are typically found close to the equator and have tremendous species diversity. Because of their proximity to the equator, rainforests generally have 12 hour days and 12 hour nights all year round and do not experience any especially cold seasonal patterns, or winter weather. Rainforest seasons tend to be classified as wet and dry seasons with an average of between 80 and 260 inches of rain per year. Temperature, in the rainforest, ranges from 68° to 93°F with humidity between 77 and 88 percent. Rainforests are consistently warm, moist environments with typically poor topsoil.

Vegetation in the rainforest is stratified, or layered. The tallest trees (often Kapok trees which can grow taller than 200 feet) make up the emergent layer which reaches above the rest of the rainforest. The emergent layer is inhabited by insects, monkeys, and birds. Next is the canopy layer, reaching up between 60 and 130 feet high. These trees have pole-like trunks with branches only spreading out at the canopy which can be 10- to 40-feet thick. The canopy is a nutrient-rich environment that is home to an incredibly diverse community. Trees of the canopy include hundreds of species of tall, broad-leafed, evergreen trees. Many of the canopy and emergent-layer trees support **epiphytes**, non-parasitic plants that live on other plants, like orchids, bromeliads, mosses, and lichens. Animals of the canopy layer include snakes, frogs, birds, bats, and monkeys. The understory layer is a dark, moist environment below the canopy; sunlight is essentially blocked from this layer by the canopy. The understory contains sparse ferns, small shrubs, and mosses. Insects, frogs, jaguars, and leopards live in the understory. The forest floor is very dark and, therefore, has almost no plants. This layer contains ants, termites, earthworms, and fungi that act as decomposers for the huge amount of organic litter that falls to the forest floor.

GRASSLAND

Grasslands are biomes that have deep-rooted grasses as their primary source of vegetation. There are several kinds of grasslands including savannahs, pampas, steppes, veldts, and prairies, depending on the location and environmental conditions of the region. The two major categories of grassland are tropical grasslands, which are hot all year round and have rainy seasons that bring torrential rains, and temperate grasslands, which have hot and cold seasons and tend to be dry and windy. Tropical grasslands, also known as savannahs, have about 25 to 60 inches of rain, all during a rainy season each year, and temperatures that range from 70° to 100°F. Temperate grasslands have an average rainfall of 10 to 30 inches per year and temperatures that range from 30° to 90°F.

The plants and animals of the world's grasslands are highly specialized in order to survive in their specific environments. Plants of the tropical grassland include, obviously, grasses and other broad-leafed plants that thrive among grasses. Scattered trees and shrubs can be found in these savannahs, but there is too little rain to support large trees or groups of trees. Animals of the African savannahs include antelope, giraffes, elephants, and lions. Kangaroos and koalas live in the North Australian savannah. Plants of the North American temperate grassland, or prairie, include many species of tall and short grasses, sunflowers, asters, and goldenrod. Animals of the prairie include beetles, crickets, eagles, prairie chickens, prairie dogs, coyotes, wolves, bobcats, and bison.

DESERT

Deserts are biomes that receive less than 10 inches of rain each year. Because they are found at all latitudes, deserts can be continuously hot (these deserts tend to be found closer to the equator like the Sahara and the Mojave), hot during the day and very cold at night (such as the Gobi and Iranian deserts), or exclusively cold (for example the Patagonian and Antarctic deserts). Hot deserts tend to have temperatures ranging from 68° to120°F. Cold deserts average from 28° to 79°F. The Antarctic desert is extremely cold most of the year with temperatures ranging from −130° to 59°F. Because of the harsh, dry conditions, biodiversity is low and those plant and animal species that live in deserts must be well adapted to their environment. Many plants (for example, cacti) and animals (for example, camels) have interesting water use and storage adaptations that are unique to desert organisms. Many desert animals are **nocturnal**, or active only during the cool of night. **Diurnal** animals (active during the day) tend to be small, may have reduced their water requirements, and many are even able to get all the water they need from their food.

Plants found in hot deserts include many species of cactus and low-growing shrubs. Spiders, beetles, snakes, mice, kangaroo rats, scorpions, porcupines, armadillos, kangaroos, dingoes, turkey vultures, bobcats, and dromedary and bactrian camels, are among the animal species that are found in hot deserts around the world. Cold deserts may have sagebrush and other low-growing bushes. Animal species found in cold deserts depend on the location of the desert. They include penguins, rhea, hawks, eagles, gerbils, pygmy armadillo, foxes, puma, antelope, and Asian tortoises.

Important Terms to Remember

Abyssal: Deepest parts of the oceans; often refers to deep sea trenches.

Benthic: The ocean floor below pelagic regions usually comprised of sand, silt, and dead organisms.

Biome: A large geographical region that contains distinctive landforms, weather patterns, and species that thrive under, and have adapted to, the local conditions.

Canopy: The top layer of a forest; this layer is dense enough to be almost continuous from one tree to the next.

Detritus: Particles or bodies of decaying organic material.

Diurnal: Animal species that are active primarily during daylight hours.

Epiphytes: Non-parasitic plants that live on other plants.

Fauna: Animal species.

Flora: Plant species.

Intertidal: Shoreline region of the ocean, where land meets the ocean.

Nocturnal: Animal species that are active primarily during nighttime hours.

Pelagic: Open water of the oceans.

Permafrost: A layer of permanently frozen soil.

Plankton: Organisms that float in the water.

Salinity: Salt concentration.

Strata: Levels of structure.

Topography: The geographic landscape of a region.

Practice Questions

Multiple-Choice

1. When a whale dies in the ocean, its carcass sinks to the _____ zone where it becomes _____.

   (A) pelagic; a producer
   (B) pelagic; a decomposer
   (C) benthic; a coral reef structure
   (D) benthic; detritus

2. In the tundra biome, trees can't grow because _____.

   (A) permafrost blocks root growth.
   (B) trees don't grow under water.
   (C) there isn't enough rain.
   (D) herbivores eat their leaves, killing the trees before they mature.

3. Forests often have layers of different types of plants and animals living at varying heights. These layers are called _____.

   (A) canopy
   (B) deciduous vegetation
   (C) strata
   (D) boreal vegetation

4. A forest environment that has trees which lose their leaves in the fall of each year is called _____.

   (A) boreal forest
   (B) deciduous forest
   (C) rainforest
   (D) taiga

5. Many animals that live in the hot desert sleep during the day and become active at night. One name for this is _____.

   (A) nighturnal
   (B) dayurnal
   (C) diurnal
   (D) nocturnal

6. Which of the following is not used to define a biome?

    (A) the continent where the biome is found
    (B) weather patterns
    (C) topography
    (D) All of these are used to define a biome.

7. Many biomes are located within specific latitudes or degrees north and south of the equator. Which two biomes can be found at both the equator and at the farthest poles of the Earth?

    (A) tundra and grassland
    (B) freshwater and temperate forest
    (C) marine and desert
    (D) taiga and marine

8. In the rainforest, the trunks of trees that comprise the canopy are often covered with vines and other nondestructive clinging plants. These plants are home for many of the rainforest's animal species. What are these nonparasitic plants called?

    (A) epiphytes
    (B) fungi
    (C) veldts
    (D) plankton

9. The Delaware River, which defines the entire boundary between Pennsylvania and New Jersey, empties into the Delaware Bay. The river encounters tidal waters at Trenton, N.J. This biome is _____.

    (A) a pelagic zone
    (B) an estuary
    (C) an epiphyte
    (D) diurnal

10. Which biome covers most of the Earth's surface?

    (A) desert
    (B) grassland
    (C) marine
    (D) freshwater

Free-Response

1. The climatogram below describes the weather patterns over one year in Iquitos, Peru. Explain what you've learned from this graph and which biome it describes.

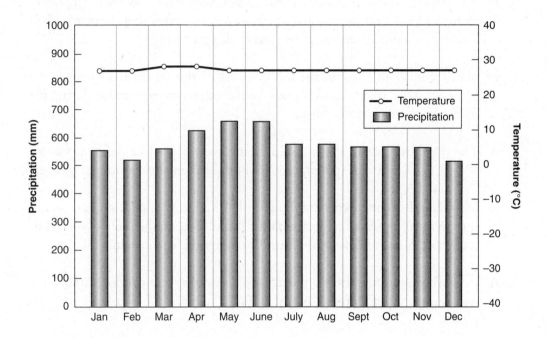

2. Describe the plants and animals that live in one biome and explain how they are well adapted to that biome.

3. Describe and explain the impact of humans on one of the Earth's biomes.

Answer Explanations

Multiple-Choice

1. **D** When a whale dies in the ocean, its carcass sinks to the benthic zone, where it becomes detritus. The pelagic zone is open ocean and the benthic zone is the ocean floor. Producers and decomposers are both living and a coral reef structure comes from coral.

2. **A** In the tundra biome, trees can't grow because permafrost blocks root growth. The tundra is not under water, its precipitation is usually in the form of snow and there's plenty of that. Herbivores may eat their leaves but not enough to kill the trees before they mature.

3. **C** Forests often have layers called strata at varying heights. The canopy is one of these layers. Deciduous vegetation includes all the trees that lose their leaves each winter and boreal vegetation includes trees with needle-like leaves.

4. **B** A temperate forest environment has trees that lose their leaves in the fall of each year. A boreal forest, or taiga, has evergreen trees that don't lose their leaves seasonally. A rainforest has leaves all year round.

5. **D** Many animals that live in the hot desert sleep during the day and become active at night; they are nocturnal. Animals that are most active during the daytime are considered diurnal.

6. **A** The continent where the biome is found is not used to define a biome. Weather patterns and topography are both elements of defining a biome.

7. **C** Marine and desert biomes can be found at both the equator and at the farthest poles of the Earth. Tundra is only found close to the North Pole, while temperate forest and taiga are found between the equator and the poles.

8. **A** Epiphytes are non-parasitic plants, often covering the trunks of trees that make up the rainforest canopy. Fungi, veldts, and plankton are not plants.

9. **B** The fresh water of the Delaware River empties into the salt water of the Delaware Bay to form an estuary. It does not form a pelagic zone, or open ocean water region. It does not form a non-parasitic vine on the trunk of a tree (an epiphyte). It isn't diurnal or most active during the daytime.

10. **C** The marine biome covers most of the Earth's surface. Desert, grassland, and freshwater are all important biomes but they do not cover nearly as much of the Earth's surface as marine biomes.

Free-Response Suggestions

1. **The climatogram below describes the weather patterns over one year in Iquitos, Peru. Explain what you've learned from this graph and which biome it describes.**

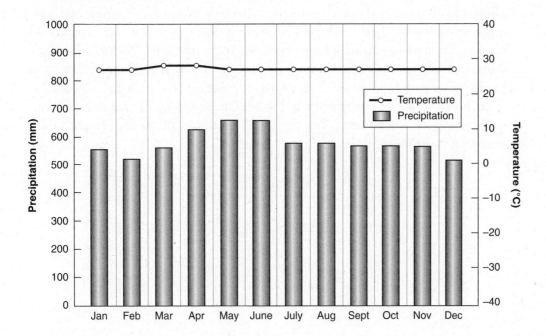

This climatogram tells me that Iquitos, Peru, has consistently warm temperatures at around 26° to 28°C, which is the equivalent of 79° to 82°F. Precipitation here averages 500 to 650 mm (20 to 25 inches) per month. These statistics indicate that Iquitos, Peru, is located in a rainforest biome.

2. **Describe the plants and animals that live in one biome and explain how they are well adapted to that biome.**

Plants of the tundra include grasses, mosses, lichens, sedges, and shrubs. These are all relatively small plants that don't require deep or extensive root systems. They must be hardy enough to survive while their water is frozen and most must be able to remain dormant during the long, harsh winter season. Animals of the tundra include species of insect that emerge at the onset of the tundra summer and quickly reproduce to thrive as a species before winter arrives again. These huge populations of insects serve as food for migrating flocks of birds, which move through marshy tundra during the summer melt. Mammals of the tundra tend to develop large stores of fat in their bodies to help them survive the less abundant winters. Many mammals of the tundra (polar bears, arctic hares, ermine, and arctic fox) have white fur, at least during the winter, to help them blend into their snowy environments.

3. **Describe and explain the impact of humans on one of the Earth's biomes.**

Humans have polluted every biome on Earth. We have dumped our garbage indiscriminately. We have depleted natural resources. Humans have impacted the forested biomes by cutting down many of the trees they contain. Humans have impacted grasslands by clearing them to build, or expand, communities. Humans have extended deserts into areas that were formerly grassland by allowing overgrazing by their herds and by cutting down trees that help to prevent erosion. We have overfished many of our marine ecosystems. We have destroyed stable ecosystems, decimated homes of many animals, and eliminated parts of food chains and food webs. Ecosystem destruction within some biomes has contributed to species extinctions and moved other species toward being endangered. Ecosystem destruction has destroyed species of plants that might otherwise have been found to be useful as medicines.

CHAPTER 16

Human Impacts

Outline for Organizing Your Notes

I. Impacts of Conservation Biology
   A. Dealing with Naturally Occurring Events
   B. Negative Human Impacts
      1. Pollution
      2. Global Warming
      3. Carbon Emissions
      4. Acidification
      5. Deforestation
      6. Habitat Destruction
      7. Endangered and Extinct Species
      8. Depletion of Fossil Fuels
   C. Positive Human Impacts
      1. Alternative Energy Sources
      2. Reduce, Reuse, Recycle
      3. Protecting Ecosystems from Non-Native Species
      4. Living Sustainably

II. Impacts of Genetic Technology
   A. Prenatal Testing and Screening
   B. Gene Therapy
   C. Cloning
   D. Pharmacogenetics
   E. DNA Profiling
      1. Predictive Medical Screening
      2. Paternity Testing
      3. Forensic Use of DNA

There are lots of reasons for learning about biology: Understanding how our bodies are arranged and how they work, learning about genetics and how we inherit traits from our parents, exploring how living things are organized and classified and how they get energy, understanding how and why adaptations happen and how they influence natural selection and evolution, and observing how ecosystems work and how species interact. Understanding our place in the world, and recognizing how we impact it, is important. Seeing what we've done right and figuring out what we've done wrong, looking at how to fix the damage we've caused, and planning how to avoid creating any new damage, is imperative.

One big difference between *Homo sapiens* and all the other organisms on Earth is that we can make conscious decisions to significantly modify our environment to meet our needs. Humans have moved into almost every terrestrial environment on Earth. We've built homes, businesses, factories, hospitals, highways, and roads. We've developed medical, architectural, nutritional, and information technologies. We've cut down forests, mined minerals, and drilled oil fields. We've burned fossil fuels, dumped garbage in the oceans, and created huge landfills. In the wake of developing a civilized and effective society, humans have left a path of habitat destruction and pollution. Just what have we done and what can we do about it now?

Impact of Conservation Biology

The scientific study of the natural world with a special interest in protecting the Earth's biodiversity is called **conservation biology**. The field of conservation biology uses approaches from natural resource management, ecology, marine biology, genetics, population biology, statistics, and economics to develop resource management plans and policy. Scientists specializing in conservation biology work to protect species and ecosystems. They identify populations of species that are at risk and habitats that are being damaged and destroyed. These scientists examine all of those criteria that are found to have an impact on biodiversity. Some threats to biodiversity happen naturally: natural disasters, geographic isolation, emigration, immigration, and inbreeding can all happen without human intervention. Other problems in biodiversity are very much man-made issues: habitat destruction, deforestation, pollution, global warming, human encroachment into pristine habitats, humans releasing invasive species into non-native habitats, and human behaviors resulting in endangered and extinct species. Let's take a look at some of the man-made conditions that are currently a concern for conservation biologists.

Pollution is the introduction of harmful contaminants, or pollutants, into the environment. There are many forms of pollution: poorly managed garbage, release of toxic chemicals, overuse and poor management of agricultural products and fertilizers, as well as poorly managed emissions from power plants, manufacturing plants, and vehicles. As the human population has grown, there has been increasing human invasion and damage in areas that were once natural and unpolluted habitats. In highly populated areas, it's common to experience light, noise, and heat pollution based on population density. Sometimes pollution is categorized based on the space being impacted: air, water, and soil. Pollution can be a product of natural processes as well as man-made conditions.

One natural process that has been aggravated by human activities is the accumulation of CO_2 in the atmosphere. While carbon dioxide is a normal by-product of cellular respiration, it is also produced in much greater quantities as a result of the burning of fossil fuels and other chemicals. The increase of CO_2, a **greenhouse gas** in the atmosphere, is contributing to **acid rain**, ocean warming and acidification, and **global warming**. Greenhouse gases also include ozone, nitrous oxide, methane, and water vapor. These gases all interact with wavelengths of light in the infrared range and, in doing so, they insulate the Earth's surface. The problem is that as greenhouse gas levels increase beyond the naturally occurring levels, they insulate even more, causing ocean warming and global warming to increase as well. The warming of oceans creates a problem with maintaining the polar ice. As polar ice melts, ocean levels rise and flood populated terrain. When carbon dioxide interacts with water, hydrogen ions are released, increasing the water's acidity and resulting in acid rain. Acidification of oceans and rain is dangerous because the living things in habitats affected by acidic water (including corals, fish, trees, and plankton) are damaged or killed when their water supply is contaminated. Global warming is an international issue because many scientists feel that it creates the potential for rising sea levels, increased size of deserts, increased extreme weather patterns, and severe climate changes.

Wasted resources and poor resource management are other areas of concern. **Deforestation**, depletion of the existing **fossil fuel** supply, and habitat destruction are all examples of wasteful use of resources and inadequate resource management. Deforestation is the result of widespread, unmanaged logging or burning of trees. Destruction of rainforests is of particular concern because these enormous and thriving forests consume huge amounts of atmospheric CO_2 during photosynthesis and are helpful in reducing levels of CO_2 as a greenhouse gas. Destruction of entire sections of forests has also been found to contribute to soil erosion and **desertification**. Without adequate reforestation efforts, many deforested areas become wastelands, habitats are destroyed, and biodiversity is lost.

Habitat destruction is a major reason for species becoming endangered or extinct. If a species no longer has the food, water, or shelter that they need, they may emigrate to find new resources, adapt to new conditions, or begin to see a severe population decline. If a species is at a very high risk of extinction in the near future, they are considered to be **endangered**. Some endangered species include giant pandas, blue whales, and orangutans. Species that no longer exist are called **extinct**; examples of extinct species include Tyrannosaurus Rex, the Dodo, and the Passenger pigeon.

Poor management of fossil fuel supplies is causing their depletion because these fuel supplies are non-renewable; they are formed by the decomposition of organisms that lived millions of years ago. Fossil fuels include oil, coal, and natural gas. The burning of these fuels both uses up our limited supply and also creates more greenhouse gases, especially carbon dioxide. One important approach to better management of energy production is the scientific development of alternative energy sources. Renewable alternatives to fossil fuels include the production of energy from wind, solar, tidal, wood, hydroelectric, nuclear, geothermal, and waste sources.

Some human impacts on environmental issues are culturally embedded, politically motivated, or difficult to change. These problems often require international agreements or legal maneuvering. There are some environmental issues that individuals can impact. Proper disposal of waste and recycling, and reducing our consumption of disposable materials, are steps we can all take to improve the health of our environment. Other suggestions include taking care to avoid releasing invasive species into non-native habitats. Plants and animals that are not native to an area can have strong impacts on other species, and on habitats where they may be able to grow without predators or controls, and where they don't belong.

As we humans live our lives, our actions clearly have an impact on our environment. Over the centuries, humans have often overused and abused their environment and taken the Earth for granted. Many people today are making an effort to change this pattern and to reduce the negative aspects of their "footprint" on the Earth. This effort encourages **sustainability**. Sustainability describes an activity that can be continued while causing only minimal impact on the environment. Sustainable living involves reprioritizing and rearranging how we live, and the cooperation of families, neighborhoods, and communities can be important. Some elements of sustainable living include use of renewable energy and green technologies, sustainable architecture, agriculture and organic gardening, sustainable transportation, and mass transit systems. People who live sustainably try to live their lives in a way that supports a healthy and mutualistic relationship with their environment.

Impacts of Genetic Technology

As we have moved forward with innovations in genetic technology, amazing and life-saving opportunities are unfolding at the same time that new questions and ethical dilemmas are appearing. Parents who have suffered the loss of a young child to Tay-Sachs can now ensure that their future embryos will not have the disease using In Vitro Fertilization (IVF) and genetic testing technologies. Other parents are provided the opportunity, through prenatal testing, to prevent other genetic or chromosomal disorders, or to choose the sex of their baby for medical reasons. These same technologies could allow still other parents an opportunity to select the baby's sex and even preferred traits, giving them a chance to design their own child. Are there any ethical dilemmas here? Well, what about parents who want to ensure that they'll have tall children? Is this right or wrong? What about the parent's motivation? Is there a difference between small parents wanting to have a child of average height and average height parents who want their child to be tall enough to become a professional basketball player? What about genes for straight teeth or 20/20 vision? If genetic screening for these characteristics was available, should it be used? Is it only cosmetic or is it an earlier intervention than braces or glasses? How much control do you think parents should have over the physical attributes of their baby?

What about modifying the genes of somebody who's already alive? Should you be able to take someone's DNA and have them cloned? The technologies of **gene therapy** and **cloning** are both commonplace today. Why is each technology used and what are the key issues around these topics? Genetic manipulation, or gene therapy, is a developing field in which genes that cause diseases are replaced with healthy genes. This therapy has recently been used with some success in treating patients with disorders including color blindness, adrenoleukodystrophy, and severe combined immune deficiency (Bubble Boy disease). Cloning is a relatively commonplace and simple procedure that involves removing the nucleus from an existing cell from one organism and replacing it with the nucleus of a cell from the organism you want to clone. Therapeutic cloning is an area of current and active research. It involves using adult cells for medical purposes, the goal being the production of regenerative tissue for patients following trauma or loss of tissue to disease. Reproductive cloning to create new individuals, like the technology that produced the first cloned sheep, Dolly, is used in farm and laboratory animals but has not been done with humans.

The science of **pharmacogenetics** examines how a person's genes influence their response to drugs. Some people are genetically programmed to be more or less reactive or sensitive to different medications. Scientists in this field expect that they will eventually be able to identify and develop medications that are specific to a patient based on their genetic profile. These matched medications should provide the most effective treatment for that patient, without the trial and error that so often guides treatment today. As more new drugs are developed and available, it becomes more likely that they will cause someone to have an adverse reaction to them. Adverse reactions to medications are very common and can range in severity from a mild rash to digestive symptoms to severe hives with breathing difficulty. Pharmacogenetics gives us the ability to match medications to patients and identify people who are likely to experience adverse reactions to medicines.

Development of individual DNA profiles is a relatively quick and easy process. They can be used diagnostically, to establish whether someone has or is a carrier of a genetic disorder, or predictively, to determine whether someone is likely to develop a disorder later on in life. They can be used to establish paternity and, in forensic investigations, to identify an individual who was present at a crime scene. One concern with the collection of these profiles is with who owns the information in them and how this very personal information can be used. If your genetic profile is used in medical research and an effective, and very expensive, treatment is developed as a result of research on your genes, do you have a right to any of the profits? If you had genetic testing because of a family history of Huntington's disease and were free of that, but the test indicated that later in life you were likely to develop cancer, or alcoholism, or a mental illness, what would you want done with that information? Would you want to know? Can you be sure your personal DNA information would remain private? Could you become the victim of medical discrimination? What do we, as a society, need to do to ensure appropriate use of genetic information?

Impact of Human Intellect

We humans are a powerful species. We've traveled throughout the Earth and into space. We've domesticated and controlled creatures much larger and more physically powerful than ourselves. We've harnessed power from the elements, from rivers, wind, and the sun. We've created, and destroyed, great civilizations. We've mapped distant planets and we've mapped our DNA. The impact of humanity on our world has been immeasurable. Yet our greatest power comes from the three-pound organ sitting inside our skulls, our brains. Our ability to learn about and understand ourselves and the world we inhabit, gives us the potential to make our own impact on the Earth both constructive and positive.

Important Terms to Remember

Acid rain: A result of carbon dioxide interaction with water, causing the release of hydrogen ions and increasing the water's acidity.

Cloning: The removal of the nucleus from an existing cell from one organism and its replacement with the nucleus of a cell from the organism you want to clone.

Conservation biology: The scientific study of the natural world with a special interest in protecting the Earth's biodiversity.

Deforestation: Cutting down or burning entire forests or large sections of forests.

Desertification: The conversion of land that was once productive into desert.

Endangered species: A species that is at very high risk of extinction in the near future.

Extinct: Species that have died off, that no longer exist.

Fossil fuel: Oil, coal, and natural gas that can be harvested and used to provide energy.

Gene therapy: A medical procedure in which genes that cause diseases are replaced with healthy genes.

Global warming: As atmospheric insulation is increased by increased greenhouse emissions the Earth's surface becomes warmer. This creates the potential for rising sea levels, increased size of deserts, increased extreme weather patterns, and severe climate changes.

Greenhouse gas: Includes carbon dioxide, ozone, nitrous oxide, methane, and water vapor. These gases all interact with wavelengths of light in the infrared range and in doing so, they insulate the Earth's surface. When greenhouse gas levels increase beyond the naturally occurring levels, they insulate too well and may cause global warming.

Pharmacogenetics: The study of how a person's genes influence their response to drugs.

Pollution: The introduction of harmful contaminants, or pollutants, into the environment.

Sustainability: An approach to living life that minimizes the impact that humans have on their environment and on the Earth.

Practice Questions

Multiple-Choice

1. Living in a city exposes people to many types of pollutants. Which of the following types of pollution is not likely to be experienced in an urban environment?

   (A) light pollution
   (B) heat pollution
   (C) noise pollution
   (D) agricultural pollution

2. Carbon dioxide is a by-product of cellular respiration and is needed by autotrophs to conduct photosynthesis. How can it also be such a problem in producing acid rain and global warming?

   (A) Burning fossil fuels creates an excess of CO_2.
   (B) Rainforests are being destroyed so their trees aren't available for photosynthesis.
   (C) Warming and acidification of oceans kills phytoplankton and seaweed.
   (D) All of the above.

3. Greenhouse gases include carbon dioxide, ozone, nitrous oxide, and _____.

   (A) methane
   (B) liquid water
   (C) oxygen
   (D) All of the above.

4. Orangutans, blue whales, and giant pandas are all at very high risk of disappearing from nature in the very near future. As a result they are considered to be a(n) _____ species.

   (A) extinct
   (B) pre-extinct
   (C) extinct in the wild
   (D) endangered

5. People who try to reduce their negative impact on the Earth by living in a way that supports a healthy relationship with their environment are said to be living _____.

   (A) predictably
   (B) carefully
   (C) sustainably
   (D) positively

6. If a child was born with a genetic disorder like adrenoleukodystrophy and his doctor offered to replace his defective gene with one that would make him healthy, this procedure is called _____.

   (A) gene therapy
   (B) therapeutic cloning
   (C) reproductive cloning
   (D) genetic screening

7. The laboratory technique of removing a mouse skin cell, extracting its nucleus and the replacing the nucleus with the nucleus from a different mouse, is called _____.

   (A) nuclear exchange
   (B) genetic screening
   (C) cloning
   (D) gene therapy

8. The study of how a person's genes influence their response to medications is called _____.

   (A) genomics
   (B) pharmacogenomics
   (C) pharmacogenetics
   (D) genetopharmacology

9. DNA profiling can be used for _____

   (A) medical treatment of an illness.
   (B) establishing paternity.
   (C) proof that someone committed a crime.
   (D) All of the above.

10. How can people reduce their negative impacts on the Earth?

   (A) by living sustainably
   (B) by learning as much as they can about their world
   (C) by not polluting
   (D) All of the above.

Free-Response

1. If an individual wants to live sustainably, what sorts of lifestyle choices might they make?

2. Describe some of the impacts that genetic technology could have on your life.

3. If a society wants its impact on the Earth to be positive and beneficial, what environmental policies might it promote?

Answer Explanations

Multiple-Choice

1. **D** Agricultural pollution is not likely to be experienced in an urban environment. Light pollution, heat pollution, and noise pollution are all experienced in cities.

2. **D** Carbon dioxide is a by-product of cellular respiration and is needed by autotrophs to conduct photosynthesis. All of these reasons create a problem in producing acid rain and global warming. Burning fossil fuels creates a lot of CO_2, rainforests are being destroyed so their trees aren't available for photosynthesis, and the warming and acidification of oceans kills phytoplankton and seaweed.

3. **A** Greenhouse gases include carbon dioxide, ozone, nitrous oxide, and methane. Liquid water is not a gas and oxygen is not a greenhouse gas.

4. **D** Orangutans, blue whales, and giant pandas are all at very high risk of disappearing from nature in the very near future. As a result, they are considered to be endangered species.

5. **C** People who try to reduce their negative impact on the Earth by living in a way that supports a healthy relationship with their environment are said to be living sustainably.

6. **A** If a child was born with a genetic disorder like adrenoleukodystrophy and his doctor offered to replace his defective gene with one that would make him healthy, this procedure is called gene therapy.

7. **C** The laboratory technique of removing a mouse skin cell, extracting its nucleus, and then replacing the nucleus with the nucleus from a different mouse, is called cloning.

8. **C** The study of how a person's genes influence their response to medications is called pharmacogenetics.

9. **B** DNA profiling can be used for establishing paternity. DNA profiling can indicate the presence of a genetic inclination to develop a disease but can't be used to treat it. Forensic DNA profiling can be used as evidence that someone was at a crime scene but other evidence is needed to prove that they committed the crime.

10. **D** People reduce their negative impacts on the Earth by living sustainably, by learning as much as they can about their world, and by not polluting.

Free-Response Suggestions

1. **If an individual wants to live sustainably, what sorts of lifestyle choices might they make?**

   An individual who wants to live sustainably will make an effort to choose to use renewable energy, transportation, and organic foods. Renewable energy could be in the form of green architecture, where their home is built from renewable products that insulate well and are safely and naturally grown. The sustainable person will make an effort to use nontoxic products for cleaning and pest control in and around their home. Their energy might come from a renewable source like solar panels or wind energy. Their food is likely to be grown without chemical pesticides and strong fertilizers. Transportation might be human powered like a bicycle, or by mass transit, or hybrid car.

2. **Describe some of the impacts that genetic technology could have on your life.**

   Genetic screening might be needed to determine if you have, or if you carry, the gene for a genetic disorder. If your parent has the autosomal dominant Huntington's disease, you can be tested and know years before it would develop in you, so that you might make life decisions accordingly. Someday, if you become a parent, your baby will be tested at birth for an ever-increasing list of genetic disorders. If you ever have a baby with a genetic disorder you'll have the option of having future embryos screened for that disorder. One day, your doctor may present you with the option of medications that are a match to your body, which will work effectively for you without risk of side effects or adverse reactions.

3. **If a society wants its impact on the Earth to be positive and beneficial, what environmental policies might it promote?**

   If a society wants to their impact on the Earth to be positive and beneficial, they will promote environmental policies that promote safety and sound resource management practice. Development of renewable energy sources, promotion of an extensive and well-designed mass transit system, encouragement of safe and non-toxic agricultural practices, and incentives to individuals who live sustainably are all sound environmental policies that would help society have a positive impact on the Earth.

Test-Taking Strategies

Taking tests can be stressful. There are strategies that can help to both reduce stress and to improve scores.

- **Know your content:** The New Jersey Biology Competency Test is designed to be focused on scientific thinking and problem solving. However, if you don't recognize and understand the concepts and terms being used it's difficult to answer the questions accurately.

- **Review your biology regularly:** Look over your class notes, review past vocabulary tests, and text book chapters, as you study for current classroom tests.

- **Try to use multiple senses when you study:** Read aloud, record yourself, rewrite, and color code your notes, or create a table of contents for your notes. Find a relationship between what you are learning about and something in your daily life. Draw, diagram, or build a model of what you are learning. Teach someone (who is willing to listen) about what you are learning.

- **Practice taking tests:** Re-use biology tests you've taken through the year, use your textbook questions and end of chapter quizzes, make up tests with separate answer keys, then take and grade them for yourself and then exchange them with your friends or study partners. Take the quizzes at the ends of each chapter of this book and review the answer suggestions, take the practice tests in this book and review the answer suggestions. The bottom line is practice taking tests. The more you acclimate yourself to the test-taking process, the less likely you are to develop "test freeze" when there is a big test in front of you.

- **While taking the test:** Be sure to read all instructions, passages, questions, and answers thoroughly. Again (because this cannot be emphasized enough), be sure to read all instructions, passages, questions, and answers thoroughly. Do not rush yourself and be aware that even if the first answer you read seems correct, that there may be even better answers later.

Answer Sheet

PRACTICE TEST 1

1 Ⓐ Ⓑ Ⓒ Ⓓ	16 Ⓐ Ⓑ Ⓒ Ⓓ	31 Ⓐ Ⓑ Ⓒ Ⓓ	46 Ⓐ Ⓑ Ⓒ Ⓓ
2 Ⓐ Ⓑ Ⓒ Ⓓ	17 Ⓐ Ⓑ Ⓒ Ⓓ	32 Ⓐ Ⓑ Ⓒ Ⓓ	47 Ⓐ Ⓑ Ⓒ Ⓓ
3 Ⓐ Ⓑ Ⓒ Ⓓ	18 Ⓐ Ⓑ Ⓒ Ⓓ	33 Ⓐ Ⓑ Ⓒ Ⓓ	48 Ⓐ Ⓑ Ⓒ Ⓓ
4 Ⓐ Ⓑ Ⓒ Ⓓ	19 Ⓐ Ⓑ Ⓒ Ⓓ	34 Ⓐ Ⓑ Ⓒ Ⓓ	49 Ⓐ Ⓑ Ⓒ Ⓓ
5 Ⓐ Ⓑ Ⓒ Ⓓ	20 Ⓐ Ⓑ Ⓒ Ⓓ	35 Ⓐ Ⓑ Ⓒ Ⓓ	50 Ⓐ Ⓑ Ⓒ Ⓓ
6 Ⓐ Ⓑ Ⓒ Ⓓ	21 Ⓐ Ⓑ Ⓒ Ⓓ	36 Ⓐ Ⓑ Ⓒ Ⓓ	51 Ⓐ Ⓑ Ⓒ Ⓓ
7 Ⓐ Ⓑ Ⓒ Ⓓ	22 Ⓐ Ⓑ Ⓒ Ⓓ	37 Ⓐ Ⓑ Ⓒ Ⓓ	52 Ⓐ Ⓑ Ⓒ Ⓓ
8 Ⓐ Ⓑ Ⓒ Ⓓ	23 Ⓐ Ⓑ Ⓒ Ⓓ	38 Ⓐ Ⓑ Ⓒ Ⓓ	53 Ⓐ Ⓑ Ⓒ Ⓓ
9 Ⓐ Ⓑ Ⓒ Ⓓ	24 Ⓐ Ⓑ Ⓒ Ⓓ	39 Ⓐ Ⓑ Ⓒ Ⓓ	54 Ⓐ Ⓑ Ⓒ Ⓓ
10 Ⓐ Ⓑ Ⓒ Ⓓ	25 Ⓐ Ⓑ Ⓒ Ⓓ	40 Ⓐ Ⓑ Ⓒ Ⓓ	55 Ⓐ Ⓑ Ⓒ Ⓓ
11 Ⓐ Ⓑ Ⓒ Ⓓ	26 Ⓐ Ⓑ Ⓒ Ⓓ	41 Ⓐ Ⓑ Ⓒ Ⓓ	56 Ⓐ Ⓑ Ⓒ Ⓓ
12 Ⓐ Ⓑ Ⓒ Ⓓ	27 Ⓐ Ⓑ Ⓒ Ⓓ	42 Ⓐ Ⓑ Ⓒ Ⓓ	57 Ⓐ Ⓑ Ⓒ Ⓓ
13 Ⓐ Ⓑ Ⓒ Ⓓ	28 Ⓐ Ⓑ Ⓒ Ⓓ	43 Ⓐ Ⓑ Ⓒ Ⓓ	58 Ⓐ Ⓑ Ⓒ Ⓓ
14 Ⓐ Ⓑ Ⓒ Ⓓ	29 Ⓐ Ⓑ Ⓒ Ⓓ	44 Ⓐ Ⓑ Ⓒ Ⓓ	59 Ⓐ Ⓑ Ⓒ Ⓓ
15 Ⓐ Ⓑ Ⓒ Ⓓ	30 Ⓐ Ⓑ Ⓒ Ⓓ	45 Ⓐ Ⓑ Ⓒ Ⓓ	60 Ⓐ Ⓑ Ⓒ Ⓓ

Practice Test 1

Multiple-Choice

Directions: Select the best answer for each question.

1. While on vacation last summer, you and your sister explored a pond that was inhabited by many frogs. By the end of your vacation, you'd both seen several living frogs that had legs in the wrong places, limb deformities, extra legs, or missing eyes. Which characteristic of life appears to be working incorrectly?

   (A) homeostasis
   (B) growth
   (C) metabolism
   (D) organization

2. Frogs can be found in many different environments. Some frogs that live in areas with freezing temperatures have an interesting adaptation. When a frog's body begins to freeze, its blood sugar levels increase significantly, acting as antifreeze. The frog's breathing and heartbeat stop. When these frogs return to non-freezing temperatures, their hearts begin to beat, they begin to breathe, they go back to normal. The production of glucose to maintain the frog's body, in spite of external surroundings, is an example of _____.

   (A) organization
   (B) metabolism
   (C) homeostasis
   (D) metamorphosis

3. Your science teacher has assigned your team to do a presentation on reported deformities in amphibian populations in Minnesota. You've found a government website that gives you the numbers of types of deformities and the species of frogs and salamanders in which they're found. What is the name for this type of data?

   (A) qualitative
   (B) quantitative
   (C) unsubstantiated
   (D) double blind

4. Your biology class has been developing experiments to assess aggressive behavior in betta, or fighting, fish. These fish can't be put in a tank together because they'll attack each other. One team has put two betta fish in bowls next to each other, one team put pictures of different color bettas next to a fish bowl, and one team put pictures of a single color betta in different sizes next to another bowl. Your classmates will be recording the responses to these different stimuli, the response is considered to be a(n) _____.

(A) control
(B) placebo
(C) independent variable
(D) dependent variable

5. You're looking at cells through a compound light microscope and you can see and identify nuclei in many of the cells. You know that you're looking at _____.

(A) an animal cell
(B) a bacterial cell
(C) a prokaryote
(D) a eukaryote

6. You've been getting leg cramps at night and your dad thinks that you may need more potassium in your diet to remedy this problem. You eat plenty of bananas and baked potatoes to get extra potassium. Which organelle will allow, or prevent, the potassium into your cells?

(A) cell membrane
(B) cell wall
(C) nucleolus
(D) lysosome

7. You have been involved in a heavy-conditioning program to prepare for your next sports season. Your muscles have had a tough workout and are sore but you need to keep building them up. Which organelles are working extra hard to make this happen?

(A) cell wall
(B) Golgi complex
(C) ribosome
(D) smooth endoplasmic reticulum

8. You're viewing a transmission electron microscope image of a piece of onion skin. Which of the following organelles should you not expect to see?

(A) vacuoles
(B) chloroplasts
(C) mitochondria
(D) centrioles

9. You've begun training with the schools cross-country team and your coach has discussed the dangers of dehydration. Which of the following will not be a concern if you don't get enough water during your run?

   (A) body temperature increasing
   (B) cells don't get the nutrients they should be getting
   (C) increased heart rate
   (D) bones become unstable

10. Your aunt has been told that because of a clogged artery in her heart, she needs to reduce her dietary intake of fats. Now she wants to eliminate all fats from her diet. Is this a good idea and why?

    (A) Yes, all fats she eats will be harmful to her because they will all add to her coronary artery blockage.
    (B) Yes, because fats are just bad for everybody.
    (C) No, because some fats are good for us, we need them and they help us out.
    (D) No, because it doesn't really matter what she eats. A heart surgeon can try to fix the blockage.

11. In biology class you've been discussing cellular metabolism and the process of building large biomolecules from smaller molecules. This process is called _____.

    (A) hydrolysis
    (B) dehydration synthesis
    (C) endocytosis
    (D) carbon fixation

12. Your older brother has been working out every day and has been taking amino acid supplements. He has a new girlfriend and wants to look "ripped." What material is he trying to build up in his body?

    (A) proteins
    (B) carbohydrates
    (C) nucleic acids
    (D) lipids

13. You just got your first pet, a goldfish. You go to a web site to read up on fish care and keep seeing references to salt water fish and you're starting to worry that your fish needs salt in its water. Fortunately, you find a good goldfish information site and discover that goldfish are freshwater fish and while a little salt may be ok, too much salt in their water can kill them. What process makes salty water dangerous to your fish?

    (A) osmosis
    (B) facilitated diffusion
    (C) endocytosis
    (D) exocytosis

14. Pickles are cucumbers that have been soaked in brine (salt water) and then stored in vinegar. The brine is considered a_____ solution because it moves water _____ the cell.

    (A) hypotonic; out of
    (B) hypertonic; out of
    (C) hypotonic; into
    (D) hypertonic; into

15. You are running in a cross-country event and you feel yourself breathing heavily. The oxygen you are gasping for is being used to metabolize the carbohydrates from the last meal you ate. This process will generate energy for you by adding _____.

    (A) disaccharide to ADP
    (B) trisaccharide to ATP
    (C) phosphate to ADP
    (D) phosphate to ATP

16. You were turning over the soil in your backyard garden and saw that you had cut an earthworm with your shovel. You did some research and you found out that earthworm cells have 36 chromosomes. If this earthworm survived and its cells undergo mitosis, one parent cell will produce _____.

    (A) two daughter cells each with 36 chromosomes
    (B) two daughter cells each with 18 chromosomes
    (C) one daughter cell each with 36 chromosomes
    (D) four daughter cells each with 18 chromosomes

17. Your work in a pathology lab has you reviewing hundreds of microscope slides every day. Today, you have to compare the number of cells on each slide that are undergoing mitosis compared to a standard, or "mitotic index." What might be indicated by a slide that has a tissue sample that has way too many mitotic cells compared to the index?

    (A) apoptosis
    (B) senescence
    (C) synapsis
    (D) immortality

18. A feral (wild) cat in your neighborhood had kittens. Your neighbor wants to trap the cat soon to have her spayed and prevent her from having more kittens. Spaying is the surgical removal of the female reproductive organs including all of the cats' eggs. These eggs were produced during the process of _____.

    (A) mitosis
    (B) meiosis
    (C) metamorphosis
    (D) synapsis

19. You are working in a medical lab and are testing for the presence of DNA in a sample. Which of the following test results will indicate that you have DNA in your sample?

    (A) the presence of uracil
    (B) the presence of deoxyribose
    (C) the presence of ribose
    (D) None of these results are conclusive for DNA.

20. A gene is to a protein as a _____ is to a _____.

    (A) pencil; paper
    (B) car; steering wheel
    (C) cookie recipe; cookie
    (D) door; house

21. The sequence of events in the production of a protein molecule is _____.

    (A) RNA → translation → replication → RNA → protein
    (B) translation → DNA → transcription →RNA → protein
    (C) DNA → transcription → RNA → translation → protein
    (D) DNA → replication → translation → RNA → protein

22. An mRNA molecule is 606 nucleotides long with an initiation and termination codon at either end. How many amino acids make up this protein?

    (A) 200
    (B) 300
    (C) 600
    (D) 1,200

23. Your friend has type O blood. His mom is type A and his dad is type B. Your friend has decided that he must be adopted. Can you help him figure this out?

    (A) Mom's genotype is $I^A i$ and dad's genotype is $I^B i$.
    (B) Mom's genotype is $I^B i$ and dad's genotype is $I^A i$.
    (C) Mom's genotype is II and dad's genotype is ii.
    (D) Listen a lot and be sympathetic, because you agree he must be adopted.

24. Your friend with the blood typing questions wants to know more. You explain that blood typing is an example of _____ with _____.

    (A) incomplete dominance; epistasis
    (B) codominance; epistasis
    (C) codominance; multiple alleles
    (D) incomplete dominance; multiple genes

25. You are at a very crowded movie and just as the movie begins, a very tall man sits down in front of you. There is nowhere else to sit and between adjusting your position to see, you find yourself wondering about the genetics behind this person's height. You know that his parents each contributed genes for height and you think about whether those genes were dominant or recessive and whether he is homozygous or heterozygous. These questions are about his _____. You also wonder whether his parents and siblings are similarly tall, now you are wondering about his _____.

    (A) transcription; translation
    (B) translation; transcription
    (C) phenotype; genotype
    (D) genotype; phenotype

26. Hemophilia is a disorder that results in problems with bleeding and blood clotting. The recessive gene for hemophilia is found on the X chromosome. Most women who carry the gene for hemophilia also have a healthy, and dominant, X chromosome so they become carriers of the disorder and it does not make them sick. Males exhibit symptoms of hemophilia when they receive this gene because their genotype is _____. This disorder is _____.

    (A) XY; metabolic
    (B) XY; sex-linked
    (C) XX; metabolic
    (D) XX; sex-linked

27. Development of a new cultivar of roses is a result of _____.

    (A) natural selection
    (B) selective breeding
    (C) sexual reproduction
    (D) all of the above

28. Which of the following is an example of a transgenic organism?

    (A) A new type of potato plant that has all the best features of its species.
    (B) A cow treated with bovine growth hormone.
    (C) A zebrafish that glows green with the fluorescence gene from a jellyfish.
    (D) all of the above

29. When it is important to identify DNA, but there is not much available to test, scientists can run a _____ and produce millions of copies of the original strand of DNA.

    (A) gel electrophoresis
    (B) polymerase chain reaction
    (C) restriction site enhancement
    (D) DNA fingerprint

30. Following an oil rig explosion and release of millions of gallons of oil into the Gulf of Mexico there were many types of attempts to clean up the massive oil spill. One type of cleanup effort relied upon the development of bacteria that could consume and break down hazardous material and oil spills. Use of these bacteria is called _____.

    (A) bioremediation
    (B) vector replacement technology
    (C) polymerase chain reaction
    (D) biogenic manipulation

31. Following photolysis, the energy available to fuel photosynthesis is no longer _____ energy, it is _____ energy.

    (A) mechanical; heat
    (B) light; heat
    (C) light; chemical
    (D) mechanical; chemical

32. Whenever aerobic cellular respiration occurs, it is known to generate a total of 40 molecules of ATP. A different number of molecules of ATP actually become available for use by the cell. The number of available ATP molecules is _____ and it is different because _____.

    (A) 30; cellular activity uses up one-quarter of any ATP generated
    (B) 36; four molecules of ATP fuel the process of respiration
    (C) 44; four molecules of ATP from photosynthesis are added to ATP generated during cellular respiration.
    (D) 48; eight molecules of ATP from photosynthesis are added to ATP generated during cellular respiration.

33. The main eukaryotic organelle used during aerobic cellular respiration is the _____.

    (A) mitochondria
    (B) chloroplast
    (C) ribosome
    (D) lysosome

34. You're looking through the powerful microscope at single-celled organisms found in samples of water from a deep sea thermal vent, a volcanic bed, and from a salt lake. They do not appear to contain a nucleus or obvious organelles. What are you most likely observing?

    (A) eubacteria cells
    (B) protist cells
    (C) archaebacteria cells
    (D) plant cells

35. You're in the field observing sea stars (commonly called star fish) moving along a tidal inlet at the shore. As you watch their star-shaped bodies, you are reminded that their symmetry is _____.

   (A) radial
   (B) bilateral
   (C) asymmetrical
   (D) lateral

36. You've been working with a scientific expedition along the Amazon and it turns out you found a species of tiny, pale, white fish that has never been seen before. As the discoverer, you get to pick the species name. Which is the correct way to write this name?

   (A) *Pygocentrus Caspericus*
   (B) *Pygocentrus caspericus*
   (C) *pygocentrus caspericus*
   (D) Pygocentrus Caspericus

37. Most organisms that are heterotrophic need to be mobile in order to find and catch their food. One heterotrophic kingdom is immobile. Which kingdom is it and how does it get its nutrients?

   (A) Protist kingdom; food comes to it and then the food is digested internally
   (B) Fungus kingdom; digests food externally and absorbs nutrients
   (C) Plant kingdom; makes its own food and absorbs nutrients
   (D) The animal kingdom; catches its food and then the food is digested internally

38. Vessels that exclusively transport blood away from the hearts of animals with closed circulation include _____.

   (A) veins
   (B) arteries
   (C) capillaries
   (D) all of the above

39. When water evaporates from the leaves of a plant into the atmosphere, it draws water up through the roots toward the leaves in a process called _____.

   (A) transpiration
   (B) transduction
   (C) adhesion
   (D) cohesion

40. Movement of water inside vascular plants travels from roots to shoots in the vessels called _____ and sugars made by the plant travel through _____.

    (A) sieve tubes; companion cells
    (B) phloem; xylem
    (C) companion cells; sieve tubes
    (D) xylem; phloem

41. Which biological molecules, present in the early Earth's oceans, combined to form the first primitive cells?

    (A) lipids
    (B) proteins
    (C) nucleic acids
    (D) all of the above

42. The red fox, grey fox, and arctic fox all evolved from a common ancestor but have variations that work for them in their habitats. These three species of fox are examples of _____.

    (A) vestigial structures
    (B) analogous structures
    (C) homologous structures
    (D) convergent evolution

43. The eye of a human serves a similar function to the eye of an octopus but their evolutionary development came from different lines of descent. Human and octopus eyes are examples of _____.

    (A) vestigial structures
    (B) analogous structures
    (C) homologous structures
    (D) convergent evolution

44. There are two tubes in the throat. The one in the front is the _____ and it is used for _____.

    (A) small intestine; water absorption
    (B) trachea; air exchange
    (C) esophagus; air exchange
    (D) trachea; movement of food

45. Which one of the following groups of words do not belong together?

    (A) prostate, bulbourethral gland, epididymus, vas deferens
    (B) esophagus, small intestine, stomach, large intestine
    (C) pancreas, ovary, trachea, lymph node
    (D) arteries, veins, capillaries, heart

46. Which of the following statements about the skeletal system is not true?

    (A) One job of the bones is blood cell production.
    (B) Bones are not living organs.
    (C) Some joints that hold bones together are immovable.
    (D) A healthy adult human body contains more than 200 bones

47. Following the eruption of Mount St. Helens in 1980, 23 square miles of surrounding forest were decimated. Some areas were so thickly covered in ash and rock, that soil was no longer available. Small plants began to inhabit this terrain and began the process of reforming soil. These small plants are considered to be _____.

    (A) primary producers
    (B) pioneer species
    (C) too small to help
    (D) apex producers

48. As new soil formed near Mount St. Helens, more species slowly returned to the area. This process is called _____ succession.

    (A) initial
    (B) secondary
    (C) preliminary
    (D) primary

49. There is a viral video that shows a lion and a crocodile fighting over a baby buffalo near a watering hole in Africa. This amazing video shows _____ competition, with two _____ (the lion and the crocodile) fighting over the baby buffalo. (Spoiler Alert: Fortunately for the little buffalo, its herd gets involved and saves it from becoming a meal).

    (A) interspecific; predators
    (B) interspecific; prey
    (C) intraspecific; predators
    (D) intraspecific; prey

50. What is unique about the organisms that live in the deepest parts of the ocean?

    (A) They are all very small or microscopic.
    (B) Their food webs are not based on photosynthetic producers.
    (C) There is nothing different about them, they're just like any other plant or animal, they just live in the ocean.
    (D) There are no living organisms in the deepest parts of the ocean.

51. The forest floor of a rainforest doesn't have much plant growth because
_____.

 (A) it's always raining.
 (B) the canopy blocks the sunlight.
 (C) its soil doesn't have enough nutrients for plants to grow.
 (D) bacteria, fungi, and worms in the soil destroy plants before they can grow.

52. In which biome is the state of New Jersey?

 (A) taiga
 (B) temperate forest
 (C) grassland
 (D) rainforest

53. Once a species becomes extinct it can only be brought back through
_____.

 (A) wildlife-based breeding programs
 (B) zoo-based breeding programs
 (C) gene therapy
 (D) Sorry, they're all gone and can't be brought back with current technology.

54. If a couple have a baby with a genetic disorder and then want to have another baby, they can ensure that the new baby will not have the genetic disorder by _____.

 (A) cloning the new baby
 (B) using IVF with genetic screening of the embryos
 (C) talking with a genetic counselor
 (D) being genetically screened themselves

55. The Japanese beetle is a very destructive insect pest that is believed to have arrived in the United States at a New Jersey plant nursery along with a shipment of iris bulbs in 1912. It is considered to be _____.

 (A) a native species since its been here for so long
 (B) an invasive species
 (C) an internationally adoptive species
 (D) a sustainable species

56. You and your lab partner designed an experiment to evaluate the impact of plant food on the growth of plants. At the end of the experiment, the five plants that received plant food were 72, 68, 78, 81, and 75 centimeters tall. This gives you an average height of _____ centimeters. The plants that did not receive food were 73, 65, 64, 74, and 73 centimeters tall for an average of _____ centimeters.

    (A) 80; 70
    (B) 70; 60
    (C) 75; 70
    (D) 72; 68

57. In the plant food experiment above (Question 56), lack of food is the _____ and plant food is the _____.

    (A) dependent variable; independent variable
    (B) independent variable; control
    (C) control; dependent variable
    (D) control; independent variable

58. In your lab you are conducting an experiment to evaluate the impact of time on bacterial population growth. You plot your experimental data on a graph. Your explanation of the decline in population at the end of the experiment involved the population reaching the maximum size that the environment could support, known by population biologists as _____.

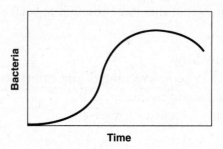

    (A) population potential
    (B) maximization
    (C) carrying capacity
    (D) maximal capacity

59. Some of your classmates have been talking about trying to maintain a lifestyle that will minimize their negative impact on the Earth. Which of the following activities would allow them the most potential impact on the environment?

    (A) growing an organic garden
    (B) using mass transit
    (C) recycling
    (D) becoming vocal and politically active about environmental issues

60. If you and your friends were to begin to live this lifestyle that would have a minimal negative impact on the Earth—planting trees, avoiding the use of fertilizers, recycling everything that can be recycled, buying environmentally friendly products, and using mass transit—the lifestyle that you would be living is considered _____.

(A) sustainable
(B) too difficult
(C) bioremediation
(D) environmental modification

Free-Response

Directions: You may use words, tables, graphs, drawings, and/or diagrams to answer the two free-response questions that follow.

1. Some of your friends have expressed an interest in making modifications to their diets. You all agree that maintaining a healthy diet is important.
   • Describe and explain what it means to have a healthy diet.
   • What biochemical molecules and elements do people need in a healthy diet?
   • Why are these compounds important to a healthy body?
   • How might someone eating a gluten-free diet (no wheat), a vegetarian diet (no meat), or a vegan diet (no animal products at all) be sure to get all the nutrients they need?

2. You recently watched a show that mentioned habitat destruction and it has made you think about where, how, and why habitat destruction has an impact.
   • How does habitat destruction happen?
   • What is the impact of habitat destruction on biodiversity?
   • How does habitat destruction cause endangered species and extinctions?
   • What can be done to minimize the negative effects of habitat destruction?

Answer Key

PRACTICE TEST 1

1. D	16. A	31. C	46. B
2. C	17. D	32. B	47. B
3. B	18. B	33. A	48. D
4. D	19. B	34. C	49. A
5. D	20. C	35. A	50. B
6. A	21. C	36. B	51. B
7. C	22. A	37. B	52. B
8. D	23. A	38. B	53. D
9. D	24. C	39. A	54. B
10. C	25. D	40. D	55. B
11. B	26. B	41. D	56. C
12. A	27. D	42. C	57. D
13. A	28. C	43. B	58. C
14. B	29. B	44. B	59. D
15. C	30. A	45. C	60. A

Answer Explanations

Multiple-Choice

1. **D** When **organization** isn't working correctly, frogs may develop with legs in the wrong places, with limb deformities, extra legs, or missing eyes. There are lots of hypotheses about the reasons for this problem including pollution, toxic exposures, and parasitic infections. A problem with homeostasis might cause the frogs to be unable to keep their hearts beating effectively or to be unable to exchange oxygen effectively when their environment changes. A growth problem would make them smaller or larger than they should be. If the problem was with metabolism, the frogs wouldn't be able to take and use nutrients from their food.

2. **C** When a body adjusts how it works in order to maintain itself it is called **homeostasis**. A change in organization would result in misplaced body parts. An alteration of metabolism would result in a change in the ability to break down or use nutrients. A change in metamorphosis would alter the transition of the frog from egg to tadpole to mature frog.

3. **B** Data that give you the numbers of types of deformities and the species in which they're found is **quantitative** data. Qualitative data would describe sensory evidence such as colors, patterns, and textures. Unsubstantiated data are information that has not been reproduced in the research of other scientists. Double blind data refer to information gathered in an experiment on humans, in which neither the scientist nor test subject knows what each test subject is being given.

4. **D** Experimental responses to different stimuli are called **dependent variable**s. Dependent variables are the results of the experiment; they depend on the independent variables. The control is the variable that hasn't been changed, that represents the normal condition. A placebo is used as a control in human clinical trials to ensure that everyone believes that they may be receiving the experimental treatment. The independent variable is the element of the experiment that is being changed and tested.

5. **D** The cell is a **eukaryote**; you know because it has a nucleus. You know it's not a bacterial cell or prokaryote because they would not have a nucleus. You don't have enough information to determine whether it is a protist, fungus, plant, or animal cell.

6. **A** The selectively permeable **cell membrane** will either allow the potassium into the cell, or not. You're an animal so you don't have cell walls. The nucleolus is involved with ribosome production and the lysosomes are involved in storing materials and breaking down waste materials, so they won't be involved in potassium entry into the cell.

7. **C** The **ribosome** is the organelle that will help you build proteins, and in turn, muscles. You're an animal so you don't have cell walls. The Golgi complex and SER function in transport, not in assembling proteins.

8. **D** You would not expect to see **centrioles** (cytoskeletal structures that look like bundles of sticks) because they are not found in plants. Vacuoles, chloroplasts, and mitochondria may all be seen in plant cells.

9. **D** The skeleton should not become any less stable because of **dehydration**. Body temperature could increase because homeostatic regulation doesn't work as well when you are dehydrated. Less body fluid will restrict delivery of nutrients to cells and as a result the heart will beat faster trying to bring more nutrients to those cells.

10. **C** No, she shouldn't eliminate all fats, or **lipids**, from her diet. There is a type of cholesterol (HDL) that helps us get rid of bad cholesterol (LDL). All of our cells are surrounded by membranes that are primarily made of a form of fat called phospholipid. Our bodies also need at least a little bit of fat to insulate us. Heart surgery is an extreme solution; often an improved and healthier diet can reduce the need for a surgical intervention.

11. **B** The process of building large biomolecules from smaller molecules is called **dehydration synthesis**, because in the process of building the larger polymer a molecule of water is lost. Hydrolysis is the disassembly of a polymer, the reverse of dehydration synthesis. Endocytosis is a form of active transport that uses energy to bring materials into a cell. Carbon fixation is the conversion of carbon from inorganic carbon dioxide to an organic carbohydrate molecule, a process that occurs during photosynthesis.

12. **A** He is trying to add to his muscle mass by increasing **proteins**, made of amino acids. Carbohydrates would give him quick energy and lipids would give him easily stored energy. Nucleic acids make up DNA and RNA, which, as nutrients, are metabolized to become nucleotides or simple molecules and elements.

13. **A** **Osmosis** makes overly salty water dangerous to fresh water fish because water will leave their cells since the surrounding fluid has more solute molecules. The fish could die of dehydration. Facilitated diffusion and active transport (endocytosis and exocytosis) won't be involved because the water's movement across a selectively permeable membrane is a problem for the fish.

14. **B** Brine is **hypertonic**; it has a higher concentration of salt molecules outside than inside the cucumber cells, so it draws water out of the cells. If it's hypertonic, water will not be pulled into the cell. If brine were hypotonic, it would cause the cucumber cells to swell with extra water.

15. **C** **ADP**, adenosine diphosphate, is a stable and low-energy molecule until a third phosphate group is added to it to form **ATP**. Adenosine triphosphate, a reactive and high-energy molecule, is the energy currency in our bodies. Oxygen helps to metabolize carbohydrates during cellular respiration; this process drives the third phosphate to bond with ADP.

16. **A** One parent cell will produce two daughter cells, each with 36 chromosomes. During the process of **mitosis**, one parent cell will produce two daughter cells identical to each other and to itself.

17. **D** Too many cells in mitosis could indicate cells growing out of control. These cells are not following the normal rules for cell reproduction and are likely indicating cancer, or cellular **immortality**. Apoptosis is programmed cell death and senescence is cellular aging that would both result in fewer mitotic cells. Synapsis is the pairing of two homologous chromosomes during meiosis, you would be seeing this in reproductive cells not in body cells.

18. **B** Egg and sperm cells are formed during **meiosis**. Mitosis produces autosomes, or body cells. Metamorphosis is the transition that occurs to change a frog or a butterfly through multiple distinct stages in its lifetime. Synapsis is an event that occurs during meiosis, which provides opportunity for variation in offspring.

19. **B** A test indicating the presence of deoxyribose will indicate that you have **DNA** in your sample. Uracil and ribose are both present in **RNA** and are not found in DNA.

20. **C** A **gene** is to a protein as a cookie recipe is to a cookie. This is because a gene gives the directions for making a protein.

21. **C** The sequence of events in the production of a protein molecule is: DNA → **Transcription** → RNA → **Translation** → Protein

22. **A** This protein will be made up of 200 **amino acids**. You know this because it had 600 nucleotides functioning as amino acid codons after you subtracted one codon each for the initiation and termination codons (606 − 6 = 600). Remember, each codon is made up of three nucleotides. So, if you divide the 600 nucleotides by the three nucleotides per codon, you find that there are 200 codons (600/3 = 200) and, therefore, 200 amino acids.

23. **A** Mom's **genotype** is $I^A i$ and dad's genotype is $I^B i$. Your friend doesn't need to worry himself about being adopted based on blood types. He inherited the recessive allele, i, from each of his parents, so his genotype is ii and his phenotype is type O blood. Blood typing is an exact science, if you receive blood that does not match your blood type it can cause a fatal reaction.

24. **C** Blood typing is an example of shared dominance, or **codominance**, and this occurs because there are multiple alleles for the same gene. Incomplete dominance would result in some intermediate phenotype rather than full expression of each allele. Epistasis is the modification, or masking, of the effect of one gene by another gene.

25. **D** When you wonder about his genetic status, about homozygous and heterozygous or about dominant and recessive genes, you are considering his **genotype**. When you think about his physical characteristics or appearance, based on his genetics, you are wondering about his **phenotype**.

26. **B** Males have the genotype XY and females have XX. Hemophilia is carried on the X chromosome and is therefore a genetic and **sex-linked** disorder. A metabolic disorder would affect the metabolism, or break down of nutrients, rather than the clotting of blood.

27. **D** Domestication of roses (and its resulting new species or **cultivars**) is a result of all three: natural selection, **selective breeding**, and sexual reproduction.

28. **C** An example of a **transgenic** organism is a zebrafish that glows green with the fluorescence gene from a jellyfish. This zebra fish has genes from a different species. The new type of potato plant that has all the best features of its species and a cow treated with bovine growth hormone are both examples of genetic manipulation of a single species. They do not have genes from other species so they are not transgenic.

29. **B** A PCR, or **polymerase chain reaction**, allows a scientist to use a small amount of DNA to produce millions of copies of the original strand of DNA. Gel electrophoresis is a procedure that provides a DNA fingerprint to match unknown to identified DNA. Restriction site enhancement would be a change to the location where enzymes can bind and cleave bacterial DNA.

30. **A** **Bioremediation** is a form of genetic engineering that develops bacteria that can consume and breakdown oil spills and hazardous material. We are not using vectors, or intermediate carriers, in this process. PCR is a laboratory procedure to amplify DNA samples. Biogenic manipulation would involve alterations to biological and genetic compounds, not the use of bacteria to clean up an oil spill.

31. **C** Before photolysis, the energy available to producers is in the form of light energy from the sun. Following photolysis, the energy available to fuel **photosynthesis** is no longer light energy, it is chemical energy, primarily in the form of ATP. Mechanical energy is not used during photosynthesis.

32. **B** The number of ATP molecules available for cellular respiration is 36. The number is different than the 40 ATP molecules generated because 4 molecules of ATP were used to fuel the process of **aerobic** cellular respiration. Individual ATP molecules are not transferred from the process of photosynthesis to the process of cellular respiration.

33. **A** The main eukaryotic organelle used during aerobic **cellular respiration** is the **mitochondria**. Chloroplasts, ribosomes, and lysosomes are not used in cellular respiration.

34. **C** Prokaryotes that are found in extreme environments like salt lakes, volcanic material, and deep sea thermal vents are likely to be ancient bacteria, or **Archaebacteria**. Few other living things survive in these harsh environments.

35. **A** Sea stars, or star fish, are round organisms that have similar structures radiating out from their center toward their multiple arms. Their **symmetry** is radial. Bilateral symmetry indicates similar structures on either side of a midline. Asymmetry describes a complete lack of mirror imagery.

36. **B** *Pygocentrus caspericus* is the correct way to write a **scientific name**. The genus name begins with an uppercase letter and the species name begins with a lowercase letter. The scientific name must be either italicized or underlined.

37. **B** The **fungus** kingdom is immobile. Fungi get their nutrients by digesting food externally and absorbing nutrients. The plant kingdom is immobile but they are autotrophs. Animals and animal-like protists are mobile heterotrophs.

38. **B** Arteries are the vessels with the singular job of transporting blood away from the hearts of animals with **closed circulation**. Capillaries are the sites of gas and nutrient exchange and veins are the vessels that return blood to the heart.

39. **A** When water evaporates from the leaves of a plant into the atmosphere, it draws water up through the roots toward the leaves in a process called **transpiration**. Transduction is a process in which genetic material is transferred from the genes of one bacterium to another. Adhesion and cohesion are properties of water that assist in the process of transpiration.

40. **D** Water and materials taken from the soil travel up through the plant inside the **xylem**. **Phloem** carries sugars produced in the photosynthetic leaves up and down through the plant, wherever nutrients are needed. Companion cells and sieve tubes are structures of the phloem that assist in moving nutrients throughout the plant.

41. **D** Lipids, proteins, and nucleic acids all combined to form the first primitive cells in the early Earth's oceans. Environmental conditions aided this process; violent lightning storms and turbulent seas helped to break and reform molecular bonds.

42. **C** The red fox, grey fox, and Arctic fox all have **homologous structures** (including different colored fur) which evolved from a common ancestor but which are adaptations that work for them in their habitats. **Vestigial structures** are body parts, such as the human appendix, that once served a purpose but that are not currently useful. Analogous structures are similar structures that came about from different lines of descent, for example, the wings of a bee and a bird. Analogous structures often appear as a result of **convergent evolution**.

43. **B** Human and octopus eyes are examples of **analogous structures**, they are similar structures that came about from different lines of descent.

44. **B** The **trachea** is one of the tubes in the throat and its job is to exchange air between the environment and the lungs. Behind the trachea is the esophagus which transports food from the mouth to the stomach. The small intestine is much further along the digestive tract.

45. **C** Pancreas, ovary, trachea, and lymph node are each from different systems. Prostate, bulbourethral gland, epididymus, and vas deferens are all structures of the male **reproductive system**. Esophagus, small intestine, stomach, and large intestine are all **digestive system** structures. Arteries, veins, capillaries, and the heart are all **cardiovascular** structures.

46. **B** Bones are living organs of the **skeletal system**. Bones do participate in blood cell production. Fixed joints are immovable, like the sutures that hold the skull together. Normal adult bodies contain 206 bones.

47. **B** **Pioneer species** are the small plants that begin to inhabit barren and soilless terrain and that initiate the process of forming soil. Any producers that begin growing here, and help with soil formation by separating rocks with their root systems, are considered pioneer species. Pioneer species tend to be small; that is not a problem for soil formation. There are no apex producers, only apex predators.

48. **D** More species slowly returned to the Mount St. Helens area in a process called **primary succession**. A **secondary succession** would occur in those areas that still had soil available for the growth of new plants. Ecologists don't refer to successions as initial or preliminary.

49. **A** This video shows an example of **interspecific** competition between two **predators**. If the competition had been intraspecific the predators would have been from the same species; here, either two lions or two crocodiles.

The organisms eating another for food are predators, while the organism that becomes food is the **prey**.

50. **B** Organisms that live in the deepest parts of the ocean are unique because their food webs are not based on photosynthetic producers. These organisms live in water too deep to receive sunlight so they use chemicals vented from below the oceans to synthesize their carbohydrates. Organisms do live in the **abyssal zone** and they are not necessarily small or microscopic.

51. **B** The forest floor of a rainforest doesn't have much plant growth because the **canopy** blocks the sunlight. While the rainforest does have frequent rain, it isn't always raining. Its soil does have enough nutrients for plants to grow. Bacteria, fungi, and worms in the soil, are decomposers; they enhance soil and do not usually destroy plants.

52. **B** The state of New Jersey is in a **temperate forest** biome. New Jersey has four distinct seasons and deciduous trees that lose their leaves in the fall. **Taiga** is a consistently colder biome than New Jersey with more evergreen trees. **Grassland** has more open plains and less wooded areas than New Jersey. **Rainforest** biomes have consistent daily rain and a steady temperature and do not have a season, autumn, in which all the trees lose their leaves.

53. **D** Once a species becomes **extinct**, they're all gone and can't be brought back using current technology.

54. **B** If a couple have a baby with a genetic disorder and then want to have another baby, they can ensure that the new baby will not have the genetic disorder by using **IVF** with genetic screening of the embryos. **Cloning** humans is illegal right now; they can't clone the new baby. Talking with a **genetic counselor** and being genetically screened themselves are useful planning tools but they won't ensure a healthy baby.

55. **B** The Japanese beetle is a very destructive insect pest that is considered to be an **invasive species**. The length of time it lives here will not qualify it to ever be considered a native species. Species are not internationally adopted into a natural habitat. Sustainability refers to a lifestyle that supports the environment.

56. **C** The 5 plants that were fed averaged approximately 75 cm (72 + 68 + 78 + 81 + 75 = 374 and 374/5 = 74.8). The 5 plants that were not given food averaged 70 cm in height (73 + 65 + 64 + 74 + 73 = 347 and 347/5 = 69.8).

57. **D** Lack of food is the **control** in this experiment and plant food is the **independent variable**. The **dependent variable** is the result of the independent variable, the height of the plants.

58. **C Carrying capacity** is known, by population biologists, as the maximum population size that the environment can support. After a population reaches carrying capacity, it is common for the population to decline before it levels out close to carrying capacity.

59. **D** All of these activities, growing an organic garden, using mass transit, and recycling, are important and do minimize an individual's negative impact on the Earth. However, speaking out and helping other people to understand how they can help has the most far-reaching potential to positively impact the environment.

60. **A** The lifestyle that involves minimizing one's negative impact on the Earth is called a **sustainable** lifestyle. People who choose this lifestyle do not find it too difficult. Bioremediation is the use of bacteria to clean up environmental spills. Changing an individual's behavior is not an environmental modification.

Free-Response Suggestions

1. A healthy diet includes the organic molecules; carbohydrates, proteins, and lipids are all crucial. A healthy diet also includes molecules like salts and water. Vitamins are organic molecules that are very important to the functioning of a healthy body. Many inorganic minerals, such as iron, calcium, potassium, and magnesium are also included in a healthy diet.

   These compounds are important to a healthy body because the cells and tissues in our bodies are continually growing, developing, repairing damage, and reproducing. We get quick energy from simple carbohydrates and storable energy from complex carbohydrates and from lipids. Proteins give us the structural materials to assemble cells and tissues, the enzymes to cause reactions, and the hormones and signaling chemicals to trigger reactions in our bodies. Vitamins and minerals work together to keep all of our tissues working effectively and to keep our systems working together.

   If someone is eating a diet that is deficient in any of the ingredients that the body needs, for example, gluten-free (no wheat), vegetarian (no meat) or vegan (no animal products at all) diets, they must be sure to get all the nutrients they need through alternative routes. A wheat/gluten-free diet may be supplemented with complex carbohydrates from other plants including quinoa, tapioca, rice, organic corn, amaranth, flaxseeds, potato, tofu, nuts, and beans. A vegetarian can get proteins via eggs, beans and rice, dairy products, legumes, peanut butter, tofu, and soy. Vegans need to take special care to take in proteins because they don't eat dairy products or eggs.

2. Habitat destruction happens when a natural habitat can no longer support the species that have lived there. Habitat destruction can be the result of a natural occurrence such as a mudslide, volcano, fire, flood, or earthquake. Human behaviors can also trigger habitat destruction through forestry practices like clear

cutting and slash-and-burn techniques of cutting down forests, through farming practices that cause desertification and soil toxicity, though bulldozing natural habitats for building space, through urban sprawl, through explosive and extensive mining practices, and through marine practices like overfishing, purse-seining, and the use of dragging nets.

All of these human behaviors cause the loss of species from their native habitats. Many members of the populations die and others emigrate, or move away. Over time, some areas recover but others never do, and species that were well-adapted cannot find a similar environment in which to thrive. Some populations begin to decline and do not have the range or appropriate habitat to reestablish themselves. This triggers a general decline in biodiversity and can cause endangered species and extinctions.

There are things that can be done to minimize the negative effects of habitat destruction. Obviously, we cannot control or prevent natural disasters. We can, though, control what we humans choose to do to our environment. On an individual level, we can make Earth-friendly choices like recycling, using mass transit, and generally living sustainably. We can choose to support merchandise and organizations that are also Earth friendly. We can communicate and share our thoughts on sustainability. We can become active in organizations that promote environmental responsibility.

Practice Test 2

Multiple-Choice

Directions: Select the best answer for each question.

1. You observe that when you run the track in gym class, you breathe faster and deeper and your heart rate increases. This is an example of which characteristic of life?

    (A) organization
    (B) homeostasis
    (C) development
    (D) movement

2. Your baby cousin seems to be growing up so fast. She's not only getting bigger, she's getting her baby teeth and she's gaining more control over her body: crawling, making sounds on purpose, and trying to stand up. Which characteristic of life is she exhibiting?

    (A) development
    (B) organization
    (C) homeostasis
    (D) metabolism

3. In order to identify organelles, you've been given cross-section photographs of interior structures of a cell. You can tell that these photographs used a technology called _____.

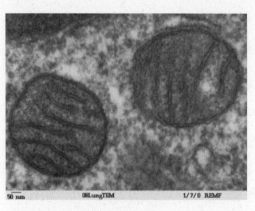

Mitochondria

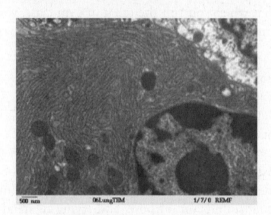

Rough Endoplasmic Reticulum

(A) SEM
(B) TEM
(C) infrared imagining
(D) digital photography

4. Your lab team is designing an experiment on the effects of plant food on plant growth. You will be keeping 10 bean plants under all the same conditions except you will be using plant food on 5 and not feeding the other 5. You will be measuring the height of each plant every day for a month. Plant growth is the _____ in this experiment.

(A) control variable
(B) independent variable
(C) dependent variable
(D) experimental variable

5. You went to the doctor for a severe sore throat, fever, and swollen glands. She swabbed your throat and said she thinks you have a bacterial infection called "strep." What kind of organism does she think is making you sick?

(A) prokaryote
(B) protist
(C) eukaryote
(D) plant

6. Your neighbor is out of town for a few weeks and she's paying you to stop over to bring in her mail and water her plants. You were invited during this time to a friend's house at the shore for a long weekend. When you got back the plants were all droopy and wilted. You quickly gave them all lots of water and in a short time they were back to looking normal and strong. Which organelle helped you get the plants looking healthy again?

(A) chloroplast
(B) nucleus
(C) vacuole
(D) RER

7. You've been given a microscope slide with many cells on it. The cells have nuclei and are all attached to each other. As far as you can tell, so far, they look very similar to each other. What can you most accurately say about these cells?

(A) They are bacterial.
(B) They are all identical.
(C) They are part of a tissue.
(D) They are part of an organ.

8. The cells on your microscope slide all appear to have a green tint, nuclei, cell walls, and large central vacuoles. What can you most accurately say about these cells?

(A) They are prokaryotic.
(B) They are autotrophic.
(C) They are decomposers.
(D) They are heterotrophic.

9. You baked a pie yesterday and some of the sticky pie filling overflowed, burning onto the bottom of the oven. Now your mom says you have to clean the oven and to be sure to wear protective gloves. She's even found you protective goggles to wear. Why is she being so cautious?

(A) Because the burnt on pie filling is really crusty and you're going to have to scrub really hard.
(B) Because the pie filling has a pH of around 2 and the acid might burn.
(C) Because the oven cleaner has a pH of around 13 and the strong base will burn.
(D) She's just overly cautious.

10. It was raining out and you and a friend were playing basketball inside the house. You knew it was a bad idea but since you were going to play anyway you looked around to see if anything needed to be moved out of the way. Your mom had a couple plants out so you stuck them in a closet for protection. Unfortunately, you forgot about them and several weeks have gone by. Your mom just found them in the closet and they are all yellow and look like they're dying. Which macromolecule were they deprived of while they were in the closet and out of the sun?

    (A) proteins
    (B) nucleic acids
    (C) lipids
    (D) carbohydrates

11. In biology class you're discussing cellular metabolism and the enzyme catalyzed process of hydrolysis; breaking large molecules down to small enough molecules to pass through cell membranes. The simple molecule that will split the larger molecule into smaller monomers is _____.

    (A) glucose
    (B) carbon dioxide
    (C) water
    (D) a fatty acid

12. Your biology class did an experiment that involved placing liver in peroxide. Fresh liver placed in peroxide produced both heat and many bubbles. Thinking about this experiment, you realize that a chemical in the liver must have triggered the reaction you have observed. Later, in a discussion about the experiment, the teacher used the word catalyst. You now know that the liver must contain _____ that break down peroxide.

    (A) enzymes
    (B) lipids
    (C) structural proteins
    (D) signaling proteins

13. When a cell needs to use energy to move molecules across the cell membrane during active transport, it gets that energy by breaking the bonds of a chemical called _____.

    (A) glucose
    (B) salt
    (C) adenosine diphosphate (ADP)
    (D) adenosine triphosphate (ATP)

14. Last week, your science teacher asked everyone to bring in a water sample today. Since you spent the weekend at the shore, you picked up a sample of water from the ocean to bring back. Several of your friends forgot about the assignment so you added some water from the water fountain in the hall and divided your sample. Now, you're looking at your sample through a microscope and while you see lots of stuff in the sample, it doesn't look like there's anything alive. What process probably killed off any living matter in your sample?

    (A) endocytosis
    (B) osmosis
    (C) pinocytosis
    (D) exocytosis

15. Many of our white blood cells fight off infection by extending their cell membrane to wrap around infectious particles. After the particles are brought inside the cells, lysosomes destroy and digest them. The process of enveloping infectious particles is called _____.

    (A) phagocytosis
    (B) exocytosis
    (C) pinocytosis
    (D) osmosis

16. Following meiosis in the cells of a whitefish, haploid cells are formed that are precursors to eggs and sperm cells. What process must these cells undergo before they become viable gametes?

    (A) mitosis
    (B) synapsis
    (C) interphase
    (D) gametogenesis

17. During which of the following processes is mitosis not apparent?

    (A) asexual reproduction
    (B) sexual reproduction
    (C) repair of cellular damage
    (D) Mitosis is used for all of these processes.

18. Which process in cellular reproduction contributes significantly to variation in a species and ultimately to evolutionary changes in a population?

    (A) crossing over
    (B) apoptosis
    (C) senescence
    (D) cytokinesis

19. In your research for a class project on pancreatic cancer, you find information about programmed cell death. You discover that cancers generally happen when the cell stops following the cellular rules about dying when they become old. Programmed cell death is called _____.

    (A) cytokinesis
    (B) senescence
    (C) apoptosis
    (D) immortality

20. You are working in a crime scene lab and are trying to isolate the different types of molecules in a body fluid sample. Which of the following would indicate to you that you've isolated RNA?

    (A) the presence of thymine
    (B) the presence of deoxyribose
    (C) the presence of uracil
    (D) None of these results are conclusive for RNA.

21. In what way is DNA replication different from transcription?

    (A) They are the same thing because both make copies of DNA.
    (B) Replication produces a single-stranded product and transcription makes a double-stranded product.
    (C) Replication produces proteins while transcription prepares the cell for mitosis.
    (D) Replication makes two copies of DNA while transcription produces a strand of RNA.

22. When you get a cut, your body needs to make proteins to repair the damaged skin. The first step in replacing these structural proteins takes place during the process of _____.

    (A) translation
    (B) transcription
    (C) gametogenesis
    (D) epistasis

23. People who have one allele for sickle cell disease and one healthy matching allele will have _____ and _____.

    (A) heterozygous superiority; vulnerability to malaria
    (B) heterozygous superiority; resistance to malaria
    (C) codominance; sickle cell disease
    (D) epistasis; no sickle cell disease

24. A friend of yours plays center on her school's basketball team. She's really long and lean. She wears glasses and braces and can never find clothes or shoes that fit her. When she went to the doctor this summer for a sports physical, the doctor heard a mild heart murmur and now wants to do some tests. What do you think the doctor is concerned about?

    (A) cystic fibrosis
    (B) Marfan syndrome
    (C) hemophilia
    (D) sickle cell disease

25. The inheritance pattern shown in this pedigree is _____.

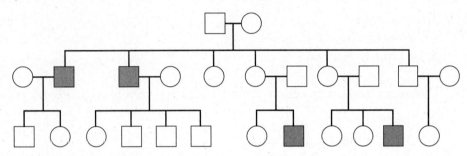

    (A) autosomal, dominant
    (B) sex linked, recessive
    (C) sex linked, dominant
    (D) polygenic, epistatic

26. Two parent pea plants are both heterozygous for two traits. There is a 9:3:3:1 ratio of these two traits among their offspring. A geneticist would call this a _____.

    (A) dihybrid cross
    (B) monohybrid cross
    (C) trihybrid cross
    (D) common cross

27. The existence of breeds of cattle with worthwhile traits, such as high milk production, is a result of _____.

    (A) sexual reproduction
    (B) natural selection
    (C) selective breeding
    (D) all of the above

28. Gene therapy is primarily used in _____.

    (A) specific groups of somatic, or body, cells.
    (B) egg and sperm cells to correct genetic disorders
    (C) somatic cells to correct multifactoral genetic disorders
    (D) all of the above

29. When a therapeutic gene is isolated, it needs to be inserted into a carrier that will transport it to its target cells in order to initiate the process of gene therapy. This carrier is also known as a _____.

    (A) hormone
    (B) cultivar
    (C) vector
    (D) somatic cell

30. The force that pulls DNA fragments across a gel plate during electrophoresis is _____.

    (A) facilitated diffusion
    (B) exocytosis
    (C) a negative electrode
    (D) a positive electrode

31. The energy needed by cells to conduct life's activities is originally generated by _____.

    (A) the cells that need the energy
    (B) the soil, water, and air surrounding living things
    (C) the sun
    (D) the food that all living things eat

32. A process that occurs in both aerobic and anaerobic respiration is _____.

    (A) photosynthesis
    (B) glycolysis
    (C) the Kreb's cycle
    (D) the light-independent reactions

33. The energy available in a molecule of ATP is stored in its _____.

    (A) adenosine molecule
    (B) bond to its first phosphate group
    (C) bond to its second phosphate group
    (D) bond to its third phosphate group

34. Animals such as frogs and butterflies transition from one appearance as a juvenile to a totally different appearance as an adult. This change is called _____.

    (A) endothermism
    (B) bipedalism
    (C) metamorphosis
    (D) conversion

35. You're walking through a neighborhood park and noticing all the different life forms around you. The cherry trees are blossoming. You know they are _____ because they have _____.

    (A) gymnosperms; fruit
    (B) gymnosperms; green leaves
    (C) angiosperms; flowers
    (D) angiosperms; green leaves

36. You've been working with a scientific expedition deep inside a cave and you have found a very active species of tiny, glowing worm that has never been seen before. As the discoverer, you get to pick the species name. Which is the correct way to write this name?

    (A) Arachnocampa Glowicus
    (B) *Arachnocampa Glowicus*
    (C) *Arachnocampa glowicus*
    (D) *arachnocampa glowicus*

37. Characteristics of water that help with its transport in complex plants include all the following except _____.

    (A) cohesion
    (B) transduction
    (C) adhesion
    (D) transpiration

38. Vessels that carry materials down the plant from the leaves to the roots are called _____ and they transport _____.

    (A) xylem; water and minerals
    (B) xylem; carbohydrates
    (C) phloem; water and minerals
    (D) phloem; carbohydrates

39. Animal classes that have a dorsal nerve cord, a notochord, and gill slits at some point in their development include _____.

    (A) birds and fish
    (B) squid and octopus
    (C) earthworms and jellyfish
    (D) insects and crustaceans

40. A transport mechanism that allows vertebrates to distribute oxygen and nutrients to all the body cells and to pick up and dispose of carbon dioxide and other waste products involves a pump and _____.

    (A) kidneys and urinary bladder
    (B) arteries, capillaries, and veins
    (C) lungs and heart
    (D) stomach, small and large intestines

41. Which of the following does not fit in with Darwin's Theory of Natural Selection?

    (A) Every organism produces offspring.
    (B) Variations are passed from parents to offspring.
    (C) Better adapted organisms will live longer and produce more offspring.
    (D) As more adapted offspring survive, variations become characteristic to the species.

42. The study of embryology relates and compares embryos of different species in an effort to establish their evolutionary relationships. Human embryos go through stages that look like the embryos of _____.

    (A) rabbits
    (B) cats
    (C) pigs
    (D) all of the above

43. As he traveled throughout the Galapagos Islands, Charles Darwin saw unique species of finch, tortoise, and iguana that were specifically adapted, from common ancestors, to the environmental conditions of the islands on which they lived. These birds and reptiles could no longer interbreed with related animals on other islands. They had changed as a result of _____.

    (A) artificial selection
    (B) adaptive radiation
    (C) geographic isolation
    (D) all of the above

44. The thyroid is considered an endocrine gland because _____

    (A) it secretes a chemical into surrounding fluid that acts in other parts of the body.
    (B) it secretes a chemical into ducts that act in other parts of the body.
    (C) it secretes a chemical that acts to end a chemical reaction.
    (D) it secretes lymphatic fluid.

45. The sequence of structures through which urine passes along its path out of the body is _____.

    (A) kidney, urethra, urinary bladder, ureter
    (B) kidney, urinary bladder, urethra, ureter
    (C) kidney, ureter, urinary bladder, urethra
    (D) kidney, urethra, ureter, urinary bladder

46. The human body system that eliminates carbon dioxide is the _____.

    (A) urinary system
    (B) respiratory system
    (C) digestive system
    (D) lymphatic system

47. In 1788, rabbits were brought to Australia to be used as a food source for European colonists. Since then, Australia has had a tremendous problem with rabbit overpopulation and the damage that rabbits have done to the country-side and to crops. This damage is primarily because there are so many more rabbits than the environment can support. The rabbits have exceeded the _____ of this ecosystem.

    (A) carrying capacity
    (B) load capacity
    (C) emigration capacity
    (D) pyramid of energy

48. Whitetail deer are the largest wild herbivores in the state of New Jersey. It is estimated that there are now around 200,000 deer in New Jersey while there were very few deer in the state 100 years ago. This population explosion has resulted in damage to the deer's habitat and, as boundaries between human activities and the forests become less distinct, many deer-related motor vehicle accidents and property damage. Movement of deer out of the forests is called _____ and movement into human-populated areas is called _____.

    (A) immigration; emigration
    (B) emigration; immigration
    (C) an abiotic factor; a biotic factor
    (D) the carrying capacity; the limiting factor

Use the following graph to answer question 49.

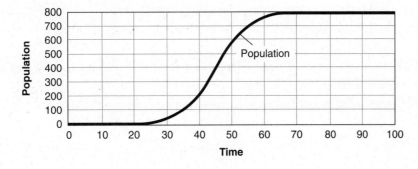

49. The graph shows a(n) _____ curve. This tells us that when the population hit 800 the environment reached its _____.

    (A) S; limiting factor
    (B) J; niche
    (C) S; carrying capacity
    (D) J; biome capacity

50. Why are so few trees found in grassland biomes?

    (A) Permafrost blocks root growth.
    (B) Root mats of grasses block the trees from developing strong root systems.
    (C) There isn't enough rain.
    (D) Herbivores eat their leaves, killing the trees before they mature.

51. Only one biome is characterized by the presence of four distinct seasons each year, summer, autumn, winter, and spring. That biome is the _____.

    (A) rainforest
    (B) taiga
    (C) desert
    (D) temperate forest

52. When carbon dioxide interacts with water in the atmosphere, acid rain is produced because _____.

    (A) hydroxide ions ($OH-$) are generated.
    (B) hydrogen ions ($H+$) are released.
    (C) carbon monoxide molecules (CO) are generated.
    (D) all of the above

53. Effective natural resource management does not include _____

    (A) replanting trees in areas where logging occurs.
    (B) building a windmill farm to harness the energy produced by the movement of air.
    (C) finding a new oil field deep under the Gulf of Mexico and building an oil rig there.
    (D) powering street lights with solar panels.

Use the following image to answer questions 54–56.

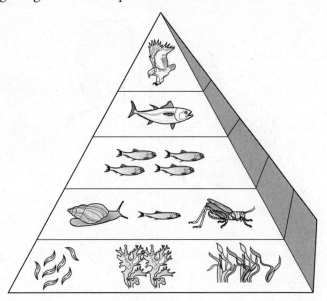

54. At each level, moving from the bottom to the top of the pyramid, the amount of available energy _____.

   (A) remains the same
   (B) keeps increasing
   (C) keeps decreasing
   (D) Cannot be determined from this image

55. At the bottom of the pyramid _____ are found, while at the top of the pyramid there are _____.

   (A) heterotrophs; autotrophs
   (B) autotrophs; heterotrophs
   (C) consumers; producers
   (D) eukaryotes; prokaryotes

56. At each level, moving from the bottom to the top of the pyramid, the number of living organisms _____.

   (A) remains the same
   (B) keeps increasing
   (C) keeps decreasing
   (D) cannot be determined from this image

Use the following image to answer questions 57–58.

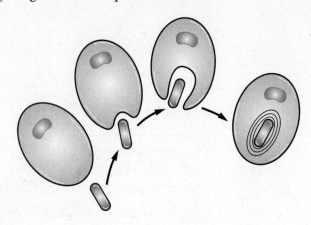

57. The type of transport represented in the image is called _____.

    (A) passive transport
    (B) exocytosis
    (C) pinocytosis
    (D) endocytosis

58. This image represents one hypothesis about the evolution of early life forms called the _____ Hypothesis.

    (A) exosymbiont
    (B) endosymbiont
    (C) mitochondrial
    (D) cell membrane

Use the following image to answer questions 59 and 60.

59. This image is called a _____ and it is a photograph of a person's _____.

    (A) karyotype; chromosomes
    (B) karyotype; chromatids
    (C) PCR; chromosomes
    (D) PCR; chromatids

60. The person represented by this image is a _____ with _____.

    (A) male; Down syndrome
    (B) male; healthy genes
    (C) female; cystic fibrosis
    (D) female; Huntington's disease

Free-Response

Directions: You may use words, tables, graphs, drawings and/or diagrams to answer the two free-response questions that follow.

1. Your family has decided that it would like to set up an aquarium and is deciding between establishing a fresh water or a salt water habitat. At the pet store, the salesperson explains that it is both more difficult and more expensive to set up and maintain a salt water tank.
   - Why is the salt water habitat more difficult and expensive to set up and maintain?
   - What kinds of organisms would inhabit each type of aquarium?
   - Why does salt content of the water matter (make sure to consider osmosis and tonicity)?

2. After learning about genetic disorders in biology class, it seems like you are hearing about genetic disorders all too often. You just heard that Brian, one of the starring members of the school basketball team, was diagnosed with the Marfan syndrome. This dominant, autosomal genetic condition causes an error in the production of a structural protein called fibrillin which acts as a glue holding parts of his body together as tightly as they should be held. Brian has been told he can no longer play contact sports because his aorta has been weakened by incorrectly made fibrillin.

   Then, the other day in history class you were discussing the impact of hemophilia on history. Your teacher explained that Queen Victoria passed this recessive, sex-linked genetic disorder onto her offspring. The ruler of early 1900s Russia Czar Nicholas, and his wife, Alexandra, had a son with hemophilia, as well. This disorder causes blood to clot much more slowly than it should and can cause serious bleeding complication after even minor injuries.
   - What are some similarities and differences in the causes of these disorders?
   - How are dominant and recessive traits inherited?
   - What is the difference between autosomal and sex-linked disorders?
   - Brian has a very tall, slender mom and an average-height dad. He also has two average-height sisters. What might a pedigree for his family look like?
   - What genotypes and phenotypes might you see in these families?
   - What might a Punnett Square for Brian's family look like?

Answer Key
PRACTICE TEXT 2

1. B	16. D	31. C	46. B
2. A	17. B	32. B	47. A
3. B	18. A	33. D	48. B
4. C	19. C	34. C	49. C
5. A	20. C	35. C	50. C
6. C	21. D	36. C	51. D
7. C	22. B	37. B	52. B
8. B	23. B	38. D	53. C
9. C	24. B	39. A	54. C
10. D	25. B	40. B	55. B
11. C	26. A	41. A	56. C
12. A	27. D	42. D	57. D
13. D	28. A	43. C	58. B
14. B	29. C	44. A	59. A
15. A	30. D	45. C	60. A

Answer Explanations

Multiple-Choice

1. **B** When you run the track in gym class, **homeostasis** causes you to breathe faster and deeper and your heart rate to increase in order to keep getting oxygen out to your cells at a steady rate. Organization would simply keep your body parts where they belong. Development refers to the changes that occur in your body as you get older. An example of metabolism would be that to nourish your body for your run on the track, you'll take in extra water and carbohydrates.

2. **A** Your baby cousin is exhibiting the characteristic called **development**; her body is changing as she gets older. Organization is seen in the predictability of the arrangement of her body; her body parts are where they belong. Homeostasis can be seen in the way her body stays in balance regardless of her external environment. Metabolism is seen in the way she takes in nutrients, gets rid of waste, and uses the energy and proteins that she takes in to stay alive.

3. **B** Cross-section photographs of organelles and other interior structures of a cell would be obtained through transmission electron microscopy, or **TEM**. Scanning electron microscopy (SEM) provides a surface view, no internal structures or cross sections; digital photography and infrared imaging are not microscopic imaging technologies.

4. **C** Plant growth is the **dependent variable** in this experiment because it is the result of the **independent variable**, plant food. The **control variable** is the group of unfed plants. All of the factors that can change in this experiment are experimental variables.

5. **A** If you have a strep infection, it is caused by a bacterium, or **prokaryote**, called *Streptococcus*. Protists and plants, as well as fungi and animals, are more complex and are **eukaryotes**.

6. **C** The **vacuole** in a plant uses water pressure to maintain the shape of cells and of the overall plant. The chloroplasts allow photosynthesis to occur but won't suddenly perk up a wilted plant. The nucleus and RER are not involved in storing water or providing water pressure to maintain the plants shape.

7. **C** These cells all appear to be connected and have similar structures and functions; they are most likely a **tissue** specimen. They have nuclei so they can't be bacteria. Cells may be very similar but are not likely to be identical and if the specimen was an organ there would be different types of tissues and, therefore, different types of cells.

8. **B** These cells are **autotrophic**. They have nuclei so they are not prokaryotes. Their green tint indicates the presence of chlorophyll and so they are not decomposers. Heterotrophic cells do not contain chlorophyll or cell walls.

9. **C** The **pH** of oven cleaner is 13 and it will burn your skin. Burnt pie filling will probably be really crusty and you will probably need to scrub a lot but you don't need goggles and protective gloves for that. Pie filling will not be a strong acid with a pH of 2. Your mom is being smart; the burn you could get from a strong base is as bad as a burn from a strong acid.

10. **D** Out of the sunlight, the plants weren't performing photosynthesis so they didn't have a chance to make and store **carbohydrates**. Plants also need proteins, nucleic acids, and lipids but they don't require the sunlight to make them available for the plant.

11. **C** Water is the molecule used in **hydrolysis** to divide polymers and dimers into monomers. Glucose, carbon dioxide, and fatty acids are not involved in breaking apart polymers.

12. **A** A **catalyst** is an **enzyme**, a protein that lowers the activation energy of a reaction, allowing chemical reactions to occur at body temperature and without mixing. You placed fresh liver in peroxide and a liver enzyme, called peroxidase, produced both heat and many oxygen bubbles as it degraded the peroxide. Lipids, structural proteins, and signaling proteins do not act as catalysts in the body.

13. **D** **ATP** is the unit of cellular energy used for active transport across cell membranes. Sugar, salt, and ADP are not used for cellular energy exchange.

14. **B** **Osmosis**, the movement of water into the sample cells in response to a lower salinity in its new, fresh water environment, has probably killed any living specimens. Endocytosis (including pinocytosis) and exocytosis involve enclosing materials in the cell's membrane and moving these materials in or out of the cell.

15. **A** **Phagocytosis** is the active transport process that involves enveloping a particle and then carrying it inside the cell to be digested. Exocytosis would involve taking the particle out of the cell. Pinocytosis involves enveloping a liquid droplet and then carrying it inside the cell to be digested. Osmosis involves the movement of water across the selectively permeable cell membrane.

16. **D** Following meiosis, the cells must undergo **gametogenesis** in order to become mature and usable egg or sperm cells. Mitosis produces identical daughter cells, not gametes. Synapsis and interphase both happen long before haploid cells are formed at the end of meiosis.

Practice Test 2

17. **B** **Mitosis** not used in the process of sexual reproduction. Mitosis is used in the process of asexual reproduction and in the repair of cellular damage.

18. **A** **Crossing over** is the time in cell reproduction when genetic information is exchanged by homologous chromosomes; this creates variation between members of a species. Apoptosis is cell death, senescence is cellular aging, and cytokinesis is cell division.

19. **C** Programmed cell death is called **apoptosis.** Cytokinesis is formation of two new cells at the end of mitosis. Senescence is normal cellular aging and immortality is the ability of a (usually cancerous) cell culture to stay alive long after the cells should have died.

20. **C** The presence of uracil would indicate to you that you've isolated **RNA.** The presence of thymine and deoxyribose both indicate the presence of DNA.

21. **D** DNA **replication** is different from **transcription** because replication makes two copies of DNA (double stranded) while transcription produces a strand of RNA (single stranded).

22. **B** The first step in replacing structural proteins in order to repair the damaged skin takes place in the nucleus during the process of **transcription.** **Translation** occurs in the cytoplasm after transcription is finished and actually joins the different amino acids to form the proteins that are needed. Gametogenesis is the formation of eggs and sperm. **Epistasis** is the masking of one gene by another.

23. **B** One healthy allele and sickle cell allele result in **heterozygous superiority** with a resistance to malaria. It wouldn't be superior, or beneficial, if the person was more vulnerable to malaria. There is no shared dominance (**codominance**) or masking of phenotypes (**epistasis**) in this situation.

24. **B** The **phenotype** being described is that of the **Marfan syndrome.** Your friend is tall and slender, with vision, dental, and cardiac issues. These can all be attributed to connective tissue weakness as a result of Marfan syndrome. **Cystic fibrosis** would present with respiratory and digestive problems and lots of thick, sticky mucous. **Hemophilia** would involve excessive bruising and bleeding and is generally seen in boys, not girls. **Sickle cell disease** would involve anemia and joint pain.

25. **B** This **pedigree** represents a disorder that's only seen in the males, it is likely to be sex linked (so it won't be autosomal or polygenic). The gene is recessive because it's not seen in some members of a generation and yet is seen in their offspring.

26. **A** A geneticist would call a $9:3:3:1$ ratio in an F_1 generation a **dihybrid** cross, indicating that both parents are heterozygous for two traits. A mono-

hybrid cross would refer to heterozygosity for a single trait and a trihybrid cross would refer to three traits. Genetically, there is no "common cross."

27. **D** Development of selected traits in farm animals like cattle is a result of all three functions: natural selection, **selective breeding**, and sexual reproduction.

28. **A** **Gene therapy** is primarily used in specific groups of somatic, or body, cells. Germ line gene therapy is still overcoming technological, medical, ethical, and legal obstacles so modification of egg and sperm cells to correct genetic disorders is relatively uncommon. Multifactoral, or multiple gene, disorders are not currently strong candidates for gene therapy.

29. **C** When a therapeutic gene is isolated, it needs to be inserted into a carrier that will transport it to its target cells in order to initiate the process of gene therapy. This carrier is known as a **vector**. **Hormones** are chemical signals that are released by one kind of tissue or organ that impact other tissues or organs. **Cultivars** are species of plants. **Somatic** cells are body cells (not gametes).

30. **D** The driving force that pulls DNA fragments across a gel plate during **electrophoresis** is a positive electrode. The negative electrode behind the wells repels the DNA fragments. Facilitated diffusion and exocytosis are both forms of cellular transport, not of movement across a gelatin plate.

31. **C** The sun originally generates the **energy** needed by cells to conduct life's activities in the form of light energy. The cells that need the energy need to get it either through manufacturing it themselves (autotrophs) or by eating something in the food chain that made it (heterotrophs). Soil, water, air, and food don't generate their own energy.

32. **B** A process that occurs in both aerobic and anaerobic respiration is **glycolysis**. The Kreb's cycle only occurs in aerobic respiration and photosynthesis and its light-independent reactions are not part of respiration at all.

33. **D** The energy available in a molecule of **ATP**, adenosine triphosphate, is stored in its bond to its third phosphate group. Adenosine and the first two phosphates form a low-energy molecule called **ADP**, adenosine diphosphate.

34. **C** The transition that frogs and butterflies go through is called a **metamorphosis**. **Endotherms** are warm-blooded animals found in the classes Aves (bird) and Mammalia. **Bipedalism** refers to walking upright on two legs. Conversion is a method of switching between units of measurement.

35. **C** Cherry trees produce both flowers and fruit (cherries) these are both characteristics that are exclusive to the division **Angiosperm**. All plants have green leaves so that would not distinguish the cherry tree classification and

the division **Gymnosperm** produces cones, rather than flowers or fruit, in its reproductive efforts.

36. **C** *Arachnocampa glowicus* is the correct way to write this **scientific name**. It is italicized and the first letter of the genus name is capitalized while the first letter of the species name is lowercase.

37. **B** Characteristics of water that help with its transport in complex plants include all the following except transduction. **Cohesion** is the attraction of water to other water molecules. **Adhesion** is the attraction of water to other kinds of molecules. **Transpiration** is when moisture is pulled through plants from the roots to the leaves. Cohesion, adhesion, and transpiration are all characteristics of water that assist in transport.

38. **D** Vessels that carry materials throughout the plant from the leaves to the shoots and to the roots are called **phloem** and they transport carbohydrates made in the leaves. Water and minerals are transported up the plants by the **xylem**.

39. **A** Animal classes that have a dorsal nerve cord, a notochord, and gill slits at some point in their development include all of the **vertebrate** subphylum, including the birds and fish. Squid, octopus, earthworms, jellyfish, insects, and crustaceans are all **invertebrates** and do not have these structures during their development.

40. **B** Arteries, capillaries, and veins are a system of tubes that use a pump, the heart, to allow vertebrates to distribute oxygen and nutrients to all the body cells and to pick up and dispose of carbon dioxide and other waste products. Kidneys and the urinary bladder filter and dispose of toxins from the blood. Lungs and heart work together to collect oxygen from the atmosphere and dispose of waste carbon dioxide to the atmosphere; they do not exchange nutrients and metabolic waste. The stomach and small and large intestines, absorb nutrients and water from food and dispose of fecal waste.

41. **A** Not every organism produces offspring. **Variations** are passed from parents to offspring. Better adapted organisms are likely to live longer and produce more offspring. As more adapted offspring survive, variations become characteristic to the species.

42. **D** Human **embryos** go through stages that look like the embryos of rabbits, cats, and pigs.

43. **C** Unique species that are specifically adapted to the environmental conditions of the islands on where they live and that can no longer interbreed with related animals on other islands, have changed as a result of **geographic isolation**. **Artificial selection** is not a cause because these changes happened naturally, without human intervention. **Adaptive radiation** is when new

variations of a species adapt to and fill a variety of ecological niches. Adaptive radiation is not the reason for these new variations, it is the result of the variations.

44. **A** The thyroid is an endocrine gland; as such, it secretes a chemical into surrounding tissue fluid that acts on other parts of the body. It does not use ducts, it does not end chemical reactions, and it does not secrete lymphatic fluid.

45. **C** Urine passes out of the body in the following sequence: kidney, ureter, urinary bladder, urethra.

46. **B** The body system that eliminates carbon dioxide is the **respiratory** system. The **urinary** system filters toxins from the blood and eliminates them as urine. The **digestive** system breaks down nutrients and then absorbs them, leaving solid metabolic waste to be disposed of by the large intestine. The **lymphatic** system circulates infection-fighting cells throughout the body and is not involved in the elimination of wastes.

47. **A** The rabbits have exceeded the **carrying capacity** of the Australian ecosystem. Load capacity, emigration capacity, and the Pyramid of Energy are not terms that have anything to do with this invasive species problem.

48. **B** Movement of deer out of the forests is called **emigration** and movement into human-populated areas is called **immigration**. Movement into or out of an environment is an issue in population density that is a biotic factor. Limiting factors in a population include movement of individuals into and out of the environment. Carrying capacity may be affected by immigration and emigration.

49. **C** The graph shows an **S curve.** It tells us when the population reaches its carrying capacity. Populations don't reach limiting factors and the graph doesn't tell us what is preventing population growth (acting as a limiting factor). A **J curve** continues to rise and never levels off and this graph does level off, so it can't be a J curve.

50. **C** There are sparse trees in **grassland** biomes because there isn't enough rain to support tree growth. Grasslands do not have **permafrost** to block root growth. Root mats of grasses are not strong enough to block the trees from developing strong root systems. Herbivores may eat their leaves, but this will not kill the trees.

51. **D** The **temperate forest** is characterized by the presence of four distinct seasons each year. The **rainforest** climate tends to stay relatively consistent throughout the year. The **taiga** has a long winter and a short summer without much seasonal transition between. The **desert** is always dry and

while temperatures may vary, the desert doesn't see much seasonal variation.

52. **B** When carbon dioxide interacts with water in the atmosphere, **acid rain** is produced because hydrogen ions (H+) are released. Hydroxide ions would contribute to an alkaline or basic change. Carbon monoxide molecules are not generally produced and do not contribute to the production of acid rain.

53. **C** Effective natural resource management does not include finding a new oil field deep under the Gulf of Mexico and building an oil rig there. Replanting trees in areas where logging occurs, building a dam across a river to harness the energy from the movement of water, and powering street lights with solar panels, are all examples of effective resource management.

54. **C** As you scan from the bottom to the top of the pyramid, the amount of available energy keeps decreasing. The **Pyramid of Energy** shows the energy-producing autotrophs at the bottom and at each successive level (**primary consumer**, **secondary consumer**, **tertiary consumer**) the heterotrophs only receive 10 percent of the energy that was available at the previous level.

55. **B** At the bottom of the pyramid **autotrophs** are found, while at the top of the pyramid there are **heterotrophs**. **Producers** are found at the bottom and **consumers** (**herbivore**, **omnivore**, **carnivore**) are the three higher tiers of the pyramid. **Eukaryotes** and **prokaryotes** can be either autotrophs or heterotrophs.

56. **C** At each level, moving from the bottom to the top of the pyramid, the number of living organisms keeps decreasing. The **Pyramid of Numbers** shows the number of organisms decreasing by approximately 90 percent from one level to the next.

57. **D** The type of active transport represented in the image is called **endocytosis**. **Passive transport** is a process that does not use energy and usually involves either small or uncharged molecules diffusing through a cell membrane, or larger molecules crossing the cell membrane through channel or gatekeeper proteins. **Pinocytosis** is a form of endocytosis in which liquid molecules are engulfed by the cell membrane. **Exocytosis** is the reverse of endocytosis involving the expulsion of materials through the cell membrane.

58. **B** This image represents one, strongly supported hypothesis about the evolution of early life forms called the **Endosymbiont Hypothesis**. This hypothesis considers the likelihood that complex cells are the result of simple, early life forms being engulfed by other, primitive, cells and ultimately those two cells working together to be an increasingly complex cell, or organism.

59. **A** This image is called a **karyotype** and it is simply a photograph of one set of a person's **chromosomes**. A **PCR**, polymerase chain reaction, is a technology that duplicates a small amount of DNA to form millions of strands of DNA for testing. **Chromatids** are paired, identical copies of DNA, joined at the centromere, which are formed to allow mitosis to produce two new identical daughter cells.

60. **A** The person represented by this image is a male with **Down Syndrome**. We can tell it's a male because of the different **sex chromosomes**, an X and a Y. A female would have two, similar, X chromosomes. We can tell he has a chromosomal disorder called Down Syndrome because he has 4 number 21 chromosomes. In order to diagnose the genetic disorders, **Cystic fibrosis** or **Huntington's disease**, we would have to test his genes rather than counting his chromosomes.

Free-Response Suggestions

1. The salt water habitat has a smaller range of acceptable conditions for its inhabitants. Salt level and temperature need to be achieved and maintained for days before adding the organisms that will live there. A fresh water habitat will balance out more rapidly; if it sits for a day, the temperature will stabilize and the water will lose much of its added chlorine. Fresh water-dwelling fish, such as goldfish and inexpensive water plants, can live in a fresh water tank. A salt water tank can be stocked with marine fish, corals, invertebrates, and ocean-dwelling plants.

    Salt content of the water matters because osmosis will cause water to pass into or out of the aquatic life based on concentration differences. Water will leave the bodies of the fish and leaves of the plants if there is a higher salt content in the water than in the aquarium's inhabitants (a hypertonic solution). Even though they live in water, they will become dehydrated and may ultimately die. If the salt concentration in the water is lower than it should be (the aquarium water is hypotonic), water will move by osmosis into the fish and plants causing them to swell. Without the toughness of a cell wall, fish cells may burst. It is important to maintain the isotonic balance in any aquarium, but it is easier to accomplish in a fresh water aquarium.

2. Marfan syndrome and hemophilia are both genetic disorders. The Marfan syndrome is an autosomal dominant disorder, while hemophilia is a recessive, sex-linked disorder. Autosomal means that the Marfan syndrome is carried on a chromosome other than X or Y and dominant means that it is expressed whenever the gene is passed to an offspring. Hemophilia is a sex-linked disorder and is, therefore, found on the X chromosome; although it is recessive it is expressed in any boy who receives it from his mother because the homologous Y chromosome does not contain genetic information to block its expression. A girl who gets the hemophilia gene usually just acts as a carrier (unless she is homozygous for the disorder) and is not affected by the bleeding disorder.

The genotypes in Brian's family are likely to be: Mom and Brian, both heterozygous (Mm) for Marfan syndrome, while Dad and the two sisters are likely to be homozygous recessive (mm). Mom and Brian both probably have the Marfan phenotype, while Dad and the sisters probably have the healthy phenotype. Queen Victoria and Czarina Alexandra would have had the genotype $X^H X^h$ (phenotype: female, healthy, hemophilia carrier). Queen Victoria's sons could have had the genotype $X^H Y$ (phenotype: healthy male) or $X^h Y$ (phenotype: male with hemophilia). Czar Nicholas and Alexandra's son, Alexei, had the genotype $X^h Y$ (phenotype: male with hemophilia).

A Punnett Square for Brian's family is likely to look like:

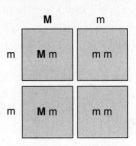

Index

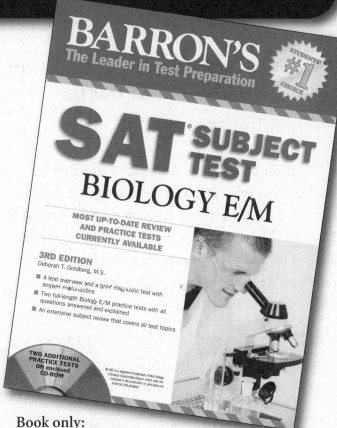